Technological Innovation in Education and Industry

Yao Tzu Li

In Collaboration With

David G. Jansson
Ernest G. Cravalho

Massachusetts Institute of Technology
Cambridge, Massachusetts

VNR VAN NOSTRAND REINHOLD COMPANY

NEW YORK   CINCINNATI   ATLANTA   DALLAS   SAN FRANCISCO
LONDON   TORONTO   MELBOURNE

Van Nostrand Reinhold Company Regional Offices:
New York Cincinnati Atlanta Dallas San Francisco
Van Nostrand Reinhold Company International Offices:
London Toronto Melbourne

Library of Congress Catalog Card Number: 79-9501
ISBN: 0-442-24754-0

Manufactured in the United States of America
Published by Van Nostrand Reinhold Company
135 West 50th Street, New York, NY 10020
Published simultaneously in Canada by Van Nostrand Reinhold Ltd.
15 14 13 12 11 10 9 8 7 6 5 4 3 2 1

Library of Congress Cataloging in Publication Data

Li, Yao Tzu.
  Technological innovation in education and industry.

  Includes index.
  1. Technological innovations. 2. Technical education. I. Jansson, David G., joint author.
II. Cravalho, Ernest G., joint author. III. Title.
T173.8.L5   607'.2   79-9501
ISBN: 0-442-24754-0

FOREWORD

The work of Y.T. Li, Ernest G. Cravalho, and David G. Jansson on *Technological Innovation in Education and Industry* is an outgrowth of their past experience as innovators and entrepreneurs as well as their work in the organization and operation of an innovation center at the Massachusetts Institute of Technology. This is one of three such centers initiated by the National Science Foundation in 1973. The other two centers are located at Carnegie-Mellon University and the University of Oregon. A fourth was added at the University of Utah in 1978.

The innovation centers had their genesis in the growing realization of the early 1970s that innovation in the United States was lagging and that the country was being increasingly deprived of the benefits that had been previously attributed to a vigorous climate for innovation in the 1950s and 1960s. It became clear that although the United States had led the world in innovation, there was a very poor understanding of the innovation process and why it did or didn't work. It appeared that if the innovation process was so vital to the economic growth and well-being of the nation, it was unfortunate that we did not understand it better and that we did not educate more young people based on that understanding. Universities traditionally viewed the elements of innovation in separate departments and schools. Entrepreneurs, who are educated in one, synthesize these elements on their own in the real world and, having done so, are hard put to explain later what they did. It seemed that there was still another role for the university: that of laying the groundwork for a discipline as well as an interdepartmental unit that combines the several elements of the innovation process in a coherent whole offering new insights and understanding. This line of thought led to a request by the National Science Foundation to Congress for appropriations to initiate three innovation centers in 1973.

In this volume, Y. T. Li and his collaborators outline their perception of the innovation process, a methodology for innovation, and a plausible

scheme by which universities can educate future innovators and entrepreneurs, while providing services that aid practicing innovators and entrepreneurs, venture-capital firms, and industry in general. It is a dis-
cerning volume that comes very close to the mark set by the National
Science Foundation in 1973. It should be of interest to a wide range of
readers.

Raymond L. Bisplinghoff
Member, National Science Board

PREFACE

The manuscript of Technological Innovation in Education and Industry was rushed out as a working draft in time for distribution at the Symposium on Innnovation and Innovation Centers, held at the Massachusetts Institute of Technology (MIT) in May 1978. At that time, the combined pressure of preparing for the symposium and hurrying to get the draft to the printer, left the author and his collaborators with little time to reflect upon its validity for the purpose for which it was intended. Five months later, however, we had developed sufficient confidence to release the draft for formal publication. In addition to many opportunities to present our work in this country, Y.T. Li gave a series of presentations and conferences in the Netherlands and the People's Republic of China on the subject of technological innovation, Ernest G. Cravalho made a presentation in Japan, and David G. Jansson spoke in a seminar on innovation in Canada. The enthusiasm of the participants in these meetings indicated that the interest in finding some stimulus to enhance the basic human instinct toward technological innovation is universal, despite diversified cultural, political, and economic backgrounds.

This book outlines a methodology that might be used to accomplish this objective, and that might also fit into the overall national movement to develop an effective climate in the United States for technological innovation. Spurring this movement, President Carter has ordered a Cabinet-level task force, headed by Secretary of Commerce Juanita M. Kreps, to give him some recommendations on this matter.

How the methodology outlined herein can be used to enhance innovation and be fitted into the nationwide movement spurred by President Carter is a question that can be examined more effectively after a review of the goal of Secretary Kreps' task force and the philosophy of innovation development. To examine this goal, let us excerpt a few lines from the keynote address* at the Symposium on Innovation and

* "The Importance of American Industrial Innovation." *Proceedings of the Symposium of Innovation and Innovation Centers.* Massachusetts Institute of Technology Innovation Center.

Innovation Centers, delivered by Dr. Frank Press, science advisor to the President, and organizer of that task force. After elaborating upon the need for innovation, Dr. Press proposed:

1. To improve long-range research supported by the federal government:

> Since World War II, there has been a pattern of research and development and commercialization in this country. It has been a successful one based on federal support of basic research conducted in the universities and federal laboratories, and to a lesser extent in industrial labs. The results of this basic research serve to underlie and lead to the applied research and development that is then phased into commercialization by the private sector. While this system has worked well, and many—including observers abroad—have praised it as being responsible for our scientific and technological leadership, it is essential that we reevaluate it and see how it might be improved. In the light of what is taking place today—the changing conditions and demands, the competition, the models of other countries that are successful—it behooves us to think deeply and freshly about this.

2. Tax incentives:

> There are many ways that altered federal policy may be able to encourage more industry investment that would lead to innovation. One way, discussed by Treasury Secretary Michael Blumenthal . . . relates to new tax incentives. Through new tax policy, we may be able to encourage industries to increase their investments. Some of this may go into capital expansion—new plants and upgraded equipment—that would have an immediate economic impact. Some of it, we would hope, might go into R&D that would have a longer-range impact. We in the Administration will pursue this.

Since the establishment of the Kreps' task force, the American economic position has worsened with the fast devaluation of the United States dollar. The "trade deficit in manufactured goods for the first half of 1978 was 14.9 billion, whereas West Germany and Japan are expected to run surpluses in manufactured goods of $49 billion and $63 billion, respectively."*

The symptoms of these setbacks, diagnosed in *Time*, are essentially the same as those reported by Press and, in fact, were known as early as the beginning of the 1970s. Thus, the National Science Foundation (NSF) was prompted to initiate the R&D incentive program and, in 1973, to fund the three innovation centers at MIT, Carnegie-Mellon University, and the University of Oregon. (The University of Utah was

* "The Innovation Recession," *Time*, October 2, 1978.

added in 1978.) The purpose of these centers was to train future entrepreneurs and innovators while they were still in college. The results of
the "experiment" were reported at the Innovation Center Symposium
and are not to be elaborated upon here.

From his experience at the MIT Innovation Center, Y.T.Li has come
to a realization of the following four basic philosophical aspects of the
innovation process:

1. The growth of industrial activities in a society are the result of the
mutual regenerative process of two separate functions, namely, R&D
activities and innovation activities. Proven innovation is the most effective vehicle for harboring R&D, while new R&D findings often provide
the stimuli for innovation. Here, the traditional "chicken and egg" concept prevails.

While the most effective way to stimulate innovation has yet to be
identified, it is sufficient to note that investment of R&D money alone
is not enough to ensure the generation of worthy innovations needed, in
turn, to regenerate support for further R&D to keep the cycle going. The
termination of the NSF RANN (Research Applied to National Need)
program, after the spending of several hundred million dollars, is a case
in point.

2. A conventional college education is an effective means for transfer
and accumulation of knowledge. While R&D work serves well to enhance product performance of a proven innovation (as noted in [1]), it is
equally effective for such accumulation of knowledge. In universities,
the process of training persons to become effective researchers and professors, thereby generating and transferring new knowledge, forms another regenerative cycle, which differs from the cycle described in (1).
At the present time, there is very little coupling between these two cycles. Thus, although increasing the R&D funding, especially in universities would help the education cycle, it is unlikely that it would benefit
the innovation cycle.

3. The preceding observations lead to the desire to find a methodology for developing skill in innovation that can be incorporated into the
traditional university curriculum for those students aspiring to be entrepreneurs and innovators. Most people accept the notion that innovators are born, not made. Yet, if the natural skill of the athlete can be
honed in training, so too can innovators be trained; indeed, coaching innovators to excel in world market competition is very much akin to the
drilling of athletic contestants for the Olympic Games.

4. The winner of a sporting event needs *willpower*, in addition to

skill, during both competition and training. The comparable behavioral element for innovation is *motivation*, which is much more complex than the willpower needed for sport. To succeed in industry, not only must the leader be motivated to take the innovation development route, he, in turn, must motivate all those workers following him. During the struggling phase of a new enterprise, motivation is sometimes more important than skill or knowledge. Thus, even if the methodology for teaching innovators is available (as hypothesized in [3]), it is still useless if students and educators are not motivated.

These four basic philosophical aspects of the innovation process served as a framework in preparing Technological Innovation in Education and Industry. An analysis of the content of the various chapters against the solution proposed in Dr. Press' talk, and the four basic philosophical aspects just described, may help the reader to get a better grip on the objectives of this work.

Chapter 1 outlines the need for innovation and calls for government support. It makes a distinction between the support of a specific innovation with an immediate market, and support for development of an innovation methodology that aims at elevating the skill of the innovative community. The former is usually considered to be a short-range goal and traditionally obtains its support from the private sector. The latter is intended for long-range social benefit and therefore deserves the support of the government.

In developing the methodology for innovation, emphasis should be placed upon its capacity for making significant incremental gains in current social needs. The "significant industrial innovations" or "quantum leaps in innovation" suggested as worthy national goals by Dr. Press are indeed a statistical phenomenon when one examines the history of innovation. In general, one tends to observe only the accumulated success of each major innovation. However, progress in development of an innovation is, in fact, made by incremental gain, which is the result of a continuous confrontation between the innovator and his financial supporter. A methodology for innovation is one that could improve the effectiveness of innovation development and the yield of the dialogue between the innovator and the supporter. This methodology could also encourage management teams of large industries to feel secure in taking a more innovative, and indeed risky, road—another concern expressed by Frank Press.

Chapter 2 outlines the logic supporting the hypothesis that innovation can be trained. The basis of this logic is the traditional engineering education, which organizes knowledge into disciplines and other structured formats to facilitate teaching, but that tends to be too rigid for innovation. In contrast, each innovator develops his own body of knowledge, which is structured upon elemental physical configurations and system behaviors. Innovation is the result of an iteration process involving the matching of configurations and perceived needs. The proposed methodology for teaching innovation, hitherto identified as the parameter analysis approach, systematizes the content of the innovator's subconscious and organizes his mental building blocks to produce a body of transferable knowledge.

Chapter 3 is a case study of a unique innovation—a C.A.T. Scanner (Computerized Axial Tomography). It illustrates the meaning of parameter analysis and reveals the modus operandi of a superb innovator who employed "gut feeling" in problem-solving.

Chapter 4 portrays the dynamic nature of a management-oriented industrial operation versus that of an innovation-oriented operation. The long-term uncertainty of the latter explains the "innovation shy" situation of many large companies. The parameter analysis methodology is considered as a remedy to this difficulty. An information flow diagram for innovation development, is proposed, which may be used to help management provide motivation to the innovation team.

Chapters 5 through 9 outline an initial attempt to develop the methodology for teaching innovation by organizing the knowledge into a sructured format. The author and his collaborators believe that by paralleling each technological discipline, a separate body of knowledge may be structured to focus upon the configuration (physical or conceptual) building blocks. For example, "frequency response," "impedance matching," and "vortex shedding" are three conceptual building blocks of the various related disciplines. When these disciplines are taught as engineering science, emphasis is placed on relating them to the fundamental elements of each discipline through mathematical modeling, such as Fourier transformation, and potential flow theory. In the innovator's mind, however, are registered all the key parameters of each useful building block with little concern directed to mathematical modeling.

Teaching innovation methodology with the parameter analysis approach is quite different from teaching engineering science. The latter is highly structured and deduction of logic is continuous, so that it can quite readily hold the student's attention. On the other hand, parameter analysis aims at developing the skill required to select a few key parameters from a collection of them. The student should be motivated to participate in that exercise inside or outside of the classroom. Thus, although it is believed that a large body of knowledge relating to parameter analysis may be structured, it should be taught through active student participation that permits students to form their own opinions first. The most exciting situation evolves where teacher and students share the burden of proof in solving new problems.

Parameter analysis as a methodology was developed at MIT by the author and his colleagues. A controversial subject, it was debated at the Symposium on Innovation and Innovation Centers with the argument centering around the question, "How do people think?" While the parameter analysis approach illustrates a thinking process, is it a process that is characteristic of most innovators and therefore capable of being developed into a methodology or, rather, does everybody think differently, so that there is no means for generalization?

Before one attempts to determine whether there is a unified thinking process for innovators, one probably can accept the observaation that a group of people exists who, by nature or training, are motivated to think analytically. They usually take the engineering-science type of education in stride. On the other hand, there is a group that claims to think intuitively and therefore is less motivated to follow the routine teaching of engineering science. What is this intuitive thinking? Can it be distilled and examined so that it can be reinforced and motivated? Thus, whether or not a unified way of thinking for innovators exists, the need for innovators is definitely present, and parameter analysis is one approach answering that need. An encouraging sign is that this methodology is now being funded for further development at MIT, and widening acceptance of its effectiveness has been expressed by the author's colleagues.

Chapter 10 illustrates an interesting exercise that provided significant motivation for students. By contrast, only a handful of students are motivated by participating in product development, even though the majority would someday like to become entrepreneurs or innovators. The primary difficulties for this latter group are their lack of experience, the long time required to realize any tangible results from product development, and hence the demand for constant attention, which conflicts with academic loading.

Chapter 10 is, as well, an illustration of the effective use of parameter analysis as a means to improve innovation, even though in this particular exercise, marketing was not an objective.

Chapter 11 illustrates the continuing struggle of innovators and entrepreneurs to gain recognition from sponsors by employing initial funding to present them with an operation with reduced risk factors. To be innovative, to have the ability to acquire capital, and to be able to complete the program, is necessary training for entrepreneurship. Parameter analysis is considered to be useful in these various steps, because it may help the innovator to make convincing presentations, to map out an effective development program, and to assist investors toward a better understanding of the new innovations presented to them.

Chapter 12 illustrates a plausible scheme with which universities can educate future innovators and entrepreneurs, while providing services that help practicing innovators and entrepreneurs, venture-capital firms, and industry in general. This scheme is conceived in recognition of needed skill, coupled with motivation for every sector of the innovative community, including the university, the faculty, and the students.

The preceding discussion illustrated that the primary goal of this book is to hypothesize a plausible process for teaching technological innovation. In a way, it reinforces the concept of improving the national long-range research process advocated by Dr. Press, which is to begin by enhancing United States industrial innovation.

In developing a methodology for teaching technological innovation, this book also emphasizes the corresponding need for motivation at every stage of the operation. The students must be motivated, the teacher must be motivated, the R&D team must be motivated, the venture capitalist must be motivated, and finally, the entrepreneur and the industrialist must be motivated so that they, in turn, can motivate the R&D team, the worker, and the consumer, and keep the economic wheel of society spinning. Accordingly, Dr. Press' proposed tax incentive is, in fact, aimed at motivating the investor and the industrialist. Indeed, all our methodology in teaching technological innovation loses its meaning unless industrialists, who are in the driver's seat of the economy, are effectively motivated. Tax incentives would help, but are they enough, and are they inflationary?

The booming decade of the sixties reminds us of our conquest of space and the resulting catalytic effect on the economy. The event is also comparable to the WPA and PWA programs President Franklin D.

Roosevelt used to revive the economy during the Great Depression of the thirties. However, such massive government expenditures might not work now, in a time of depleted resources and restrictions imposed by environmental concerns.

It may seem to be beyond the scope of this book to venture into diplomatic problems, however, in terms of overall national motivation, the two-way market potential provided by the People's Republic of China should not be overlooked. It is encouraging to note that Dr. Press visited China not long ago to open up an active dialogue for training significant numbers of Chinese students and engineers. In addition, in the absence of the radical element—the Gang of Four—the Chinese government is advocating the adaptation of Western technology, along with pseudo-Western management and marketing systems, in order to modernize the country. Needless to say, their goal is to achieve Western sophistication in technology, rather than maintain a position as a second-rate industrial nation, and they are willing to pay for it. With China's 900 million people and well-organized social system, this objective represents a huge human resource (along with its oil reserves) to activate the economic systems of the Western world. With adequate preparation, many industries in the United States could be motivated once again to become involved, even though they would have to compete with Japan and Europe, which were there first. It will be interesting to see whether overall innovation, in an international marketing game, can compensate for our diminishing lead in private industrial innovation.

Y.T. LI
ERNEST G. CRAVALHO
DAVID G. JANSSON

ACKNOWLEDGMENTS

People often wonder whether innovation is strictly the result of individual talent or whether it has sufficient latitude to be enhanced by certain forms of education. We believe that the latter is the case, akin to athletic achievement, which can be improved by coaching and exercise. The material in this working draft represents an initial attempt in that direction. While it is still rough and its premises, to a certain extent, debatable, it may serve as a lure to attract wise counsel as according to an old Chinese saying, "A brick is thrown to induce the jade."

The author and his collaborators are greatly indebted to the National Science Foundation for supporting the innovation center experiment and for making possible the symposium on Innovation and Innovation Centers, where the need for teaching technological innovation is to be examined.

The authors express appreciation to the National Bureau of Standards and the Department of Energy for their support in the development of energy-related invention evaluation methodology to add to the depth of this program.

The encouragement of Robert Colton of N.S.F., George Lewett of N.B.S., and Patrick Donohoe of D.O.E., as well as the participation of Raymond Bisplinghoff, Wayne Brown, Howard McMahon, and Ray Stata, who discussed this work at the symposium, are much appreciated.

The author and his collaborators also wish to acknowledge the advice and cooperation of their many associates at M.I.T., in particular, Professors Walter Rosenblith, Thomas Jones, James Bruce, Rene Miller, Wallace Vander Velde, Theodore Pian, S.Y. Lee, the members of the innovation education council, and the members of the innovation center advisory board.

Appreciation also to Professor T.Y. Lin of the University of California and Professor Larry Ho of Harvard.

Finally, credit and appreciation go to Virginia Peltier for her effort and skill in editing and coordinating rough material into a neatly packaged manuscript in time for the Symposium and to Frantiska Frolik for typing our many drafts.

Yao Tzu Li

CONTENTS

1
THE NEED FOR
INNOVATION

1. AN HISTORICAL NOTE ON INNOVATION

Human beings distinguish themselves as the only creatures with the ability to innovate: It is a form of survival instinct comparable to that which guides a pigeon homeward and instructs bees to colonize—those uncanny abilities of various species which defy imagination. If man-made innovations are measured by technical achievement with existing tools, then it is difficult to say whether or not there has been any significant gain in the ability to innovate over time. Although the tools have changed with the continuous advances made in science and technology, it is unclear whether contemporary man's instinctive ability to innovate differs from that displayed by earlier civilizations. For instance, to this day, people have not quite figured out how the Egyptians designed and built their great pyramids from huge stone blocks weighing many tons each. The combination of size, precision, and intricacy is a wonder of accomplishment (if measured by the tools we think they had then: hammers, chisels, pulleys, and tackles). As a technical innovation, how can we say the pyramid is insignificant in comparison with its modern day counterpart, such as the Apollo lunar module? Indeed, we know it is possible to build a pyramid with modern tools, while it was certainly beyond ancient people to reach for the moon. Imagine, though, the task of building a pyramid today using only simple blocks and tackles.

Aside from giving the ancient Egyptians credit for being very innovative in constructing the great pyramids, it is also interesting to note that the construction of the pyramids satisfied peculiar social needs. According to some anthropologists, the construction of the pyramids was

1

not just for the glorification of the Pharaohs, but, to a great extent, it was for the satisfaction of the people who participated in their construction for fun, on the order of attending olympic games. The theory is that in ancient Egypt the annual flooding of the Nile was so predictable that the major task of agriculture took place along its banks. All people had to do was to sow the seeds and reap the harvest, which represented two periods of a few weeks of hard work each year with long periods of idle time in between. Thus, building pyramids was a very effective scheme for diverting the people's excess energy; certainly a better diversion than waging war, though they did that quite often, as well.

Along a similar line of reasoning, a social need not much different from that served by the pyramids of ancient Egypt, was filled by the space program of the 1960s in the U.S. Immediately after the recession of 1957, and the firing of the first Russian Sputnik, President Kennedy initiated a 40 billion dollar-space program which produced the economic boom of the following decade, in addition to satisfying national pride by achieving the state goal of putting men on the moon within ten years.

The preceding examples illustrate that innovation is the basic human instinct to strive for effective ways to utilize the environment and, like the instincts of all creatures, provides the species with the ability to survive in the biosphere of the earth. While survival depends upon the effective use of instinct at the moment of need, the continuous exercising of instinct is, in itself, a necessary condition to maintain the well-being of each species. Thus, the pyramids and the Apollo project may be viewed either as outpourings of innovation or as an appropriate exercise to keep people mentally fit, paralleling, or serving as a precursor to, the prosperity of the society.

If the economic boom of the sixties was the result of the "outpouring of innovation" in association with the space program, what was the cause of the recession that followed? With the energy constraints which now loom over our heads will the recession grow even worse, or should we attempt to find a solution through some activation of the innovation process?

2. THE CAUSES OF THE RECENT RECESSION

The present recession started in 1968, when the moon exploration program was completed, and the entire space program was cut by more than one half. At the same time, the "Route 128 phenomenon," which had grown out of the postwar boom, collapsed like a giant balloon that had suddenly sprung a leak. Many engineers and others were out of jobs for several months, and some never returned to the professions for which they had been trained. Understandably, those who suffered most were the proposal writers who, previously in great demand, were the first to go.

Paralleling the cutback in aerospace research (including the ill-fated, supersonic transport program) venture capital suddenly dried up, largely due to the yo-yo effect created by the venture capitalists themselves. Venture capital had been stimulated by the fantastic growth of a few electronic and computer firms, which led to an unrealistic, bandwagon effect. Prices for some growth stock rose to unrealistic levels of 40-80 times the earnings, and many investors did very well. The most sensational success was probably achieved by the American Research and Development Corp., organized by General Doriot, the pioneer in funding for technical innovation. Their masterpiece was the investment of 70,000 dollars in 1957 in exchange for a control interest in the newly founded Digital Equipment Corporation. While many of the fly-by-night organizations fell by the wayside, Digital and several others, with sound foundations in engineering and management, grew rapidly through this difficult period. *Fortune* reports that Digital reached an annual volume of 736.3 million dollars (1976 figure), which represents an average annual growth of 40%, starting with moderate sales of 5.7 million dollars in 1964.

The so-called fly-by-night operations, founded by certain speculation specialists in the late sixties, knew the public to be easily attracted by the sensational success of a few but initially ignorant of the failure of many. These speculators encouraged the technical community to come forth with new companies by offering them attractive, initial investments. The plan was to get these tender, inexperienced organizations to establish plush setups and indulge in grandiose dreams of "going public" in a year or two, at which point a skillfully prepared "red herring" was to be introduced to offer shares to the public at four or more times the initial investment. In this fashion, the specialists not only made a "fast buck" but were able to enjoy the benefits of "capital gain," which the federal government taxes at a lower rate than ordinary income. Because of this kind of inducement, many shaky operations were pushed beyond reality toward inevitable collapse. Collectively, these events caused the public to lose confidence in the stock market in general.

At this time, the call for innovation in the U.S. rose from an occasional murmur to incessant, background hum. One familiar theme heard in numerous seminars at that time was the loss of the U.S. lead in innovation to other nations, such as Japan and West Germany. Both casual observation and hard statistics support this trend: On U.S. highways, one can easily see the year by year growth in numbers of Volkswagens, Audis, Toyotas, Datsuns, and many other foreign cars; in stores, Sony televisions and Japanese cameras represent quality merchandise, in contrast to the mechanized toys—some still retained the U.S. beer can decor on the inside—which flooded the U.S. market during the postwar years. Even more amazing is the dominant position of Japanese steel and shipbuilding industries which are built on raw materials shipped from other countries, including the U.S.

Nineteen seventy-two represents a year when U.S. industry took a backseat to its agricultural potential in foreign trade. This was also the year when the Japanese industrial giant dreamed about overtaking the U.S. on various fronts in the near future—a dream which was cushioned, however, by the oil embargo of 1973, which underscored the fact that, in terms of resources, Japan is certainly more vulnerable than the U.S. The Arab oil embargo of 1973 was a major historical event. It was triggered by the conflict between the Arab states and Israel and had undertones of "blackmailing" the Western world for its pro-Israeli stand. It may however, be remembered favorably in the future as a timely warning about the world's dependence on a limited supply of fossil fuel, which paved the way for affirmative energy policy of the Carter administration.

Thus, in the last ten years since 1968, the U.S. economy has suffered sequential setbacks, starting with drastic reduction in massive government spending on the aerospace program, collapse of fly-by-night venture capital operations, the ripening of foreign industrial competition and, finally, the energy crunch. As a result, we see sustained inflation and a high rate of unemployment, along with other forms of social unrest. In the face of these difficulties, the general consensus that we need more innovation has been readily reached. To the American people, it represents, in part, regret for lost leadership in technological innovation through the "technology transfer" process and, in part, breast-beating, with the hope that the inherent vitality which characterized U.S. industry in the past might be rekindled. However, despite enthusiasm for the principles underlying innovation, there is no consensus as to what innovation is and how it may be utilized to solve economic and social problems.

3. SOME VIEWPOINTS ON THE DECLINE OF INNOVATION IN THE U.S.

Big business tends to blame the decline in innovation on the government, charging it with failing to provide incentives through taxation and for ever-increasing regulations, such as those of the Environmental Protection Agency (EPA), Occupational Safety and Health Administration (OSHA), etc. American business also looks with envy at the well organized Japanese company-worker relationship, focusing especially on the workers' loyalty to the company, which seems to stimulate innovation, while tending to shrug off the cradle-to-grave responsibility the company assumes toward the worker. To some industrial managements, innovation is a commodity which can be purchased at a price, like capital equipment, therefore, they blame the loss of U.S. leadership in industrial performance on the decline in profits, which has retarded investment for the purpose of innovation and more efficient equipment. On the other hand, some sociologists seem to be concerned that the individual innovator is a breed threatened with extinction along with whales and elephants: Modern technology is so sophisticated, requiring elaborate equipment and skilled technicians that no single innovator, utilizing basement and garage methodology, has the resources to compete.

In contrast to these opinions, another school of thought tends to attribute to the individual innovator or entrepreneur a considerable contribution to the economy, and, indeed, data may be cited to show the large percentage of patent applications which are filed by individual inventors. The goal of this group, including many legislators and government officials on both the state and federal level, was to find ways to help small technological or growth-oriented business get started by providing fledgling entrepreneurs with access to managerial skills, with particular emphasis on securing venture capital, bank loans, and financial assistance from government backed operations like the Small Business Administration (SBA).

Coinciding with the promotion of technological innovation in the U.S., several European nations have adopted similar schemes; many follow the pattern of the British National Research and Development Commission (N.R.D.C.) established in 1957. The significant difference between the N.R.D.C. and the various U.S., state-run, technological development organizations lies in the N.R.D.C.'s internally funded product development program, which generates new products with patent protection or well defined know-how. In many respects, the function of N.R.D.C. is similar to that of privately owned organizations, such as Arthur D. Little, the Battelle Institute, and the Stanford Research Institute in the U.S.

Thus, the concept of innovation development for sale, as practiced by Arthur D. Little and the N.R.D.C., by the prudent and innovative venture capital investment houses, such as the American Research and Development Corporation; and, of course, by numerous research centers affiliated with large industrial complexes, such as Bell Laboratories, was in existence prior to the quadruple blow (the backlash of venture capital investment, cutback in government sponsored research programs, foreign competition, and energy constraints) producing the economic upheaval of this past decade. If these various groups did not help to provide the innovation needed to forestall economic upheaval, how could any new, state-run, innovation organization be more effective? This question leads to another even more fundamental: Whether the components of the inherent triad of our society at large—namely technology, resources, and social needs—are fundamentally out of equilibrium to the extent that innovation, itself, might become self-defeating, as some environmentalists have suspected.

4. THE GROWTH OF THE "CONSUMER-CENTERED" SOCIETY

To examine this question in its proper perspective, let us treat the half century between 1920 to 1970 as the "consumer-centered period" and the period from 1970 until the time (hopefully in the not too distant future) when our dependence on fossil fuels will be alleviated, as the "energy constraint period." The consumer-centered period was initiated by Henry Ford Senior when he introduced the mass production method together with a policy of treating his workers as consumers by paying them so that their earnings were high enough to make them the owners of their output. As a result of that policy, production zoomed in order to supply fast growing consumer needs and changed the economy and the concept of capitalism completely. Prior to the consumer-centered period, a capitalist was indeed a member of the exploiting class, as defined by Marx, while during the period, the capitalist became an agent or manager. As such, he fulfilled the needs of the people, while deriving significant earnings not from outrageously high profits but from a production quantity much higher than that ever dreamed of by his predecessor, the exploitive industrialist. His profit was fair reward for his leadership skill and the risks that he undertook by way of investment. His success depended almost exclusively on his ability to serve his customers—the consuming public, including his workers. In this goal of serving the customer, U.S. industry succeeded magnificently.

Over these last 50 years, consumers' needs have been satisfied one by one according to human biological necessities and the extensions of their anatomical functions. The satisfaction of the American peoples' biological necessities, including the supply of food, clothing, shelter, and sanitation facilities, has long surpassed the comfort margin to border on a level of universal luxury by world standards. (This does not mean that there is no squalor in the U.S.)

The extension of anatomical functions began with the introduction of various machine tools and power supply systems designed to expand the capabilities of the hands; the introduction of automobiles and airplanes to broaden the function of the legs; the introduction of radio, audio systems, and the telephone to expand the use of the ears; the perfection of photography, movie, and T.V. to add range to the eyes; and, finally, in the last 20 years, the grand entry of the computer, which extended brain activity to a level just short of invading the individual's sense of mastery. In addition, technology was developed for medicine, education, business, defense, recreation, and leisure.

In taking such an inventory by itemizing the biological needs and anatomical functions of humans, one begins to be amazed at how much has been developed over the past 50 years and how little seems to remain that really requires innovation. If just to satisfy the American ego, it might be interesting to note that out of the major innovations listed above, Americans are responsible either for the primary invention or the major development of each item—perhaps due to the abundance of resources and density of population. Indeed, if humans had more major biological functions to be enriched, the methods of enhancement would probably have been invented by Americans! As things stood, it was logical for Americans to move toward outer space and precede the Russians in putting a man on the moon.

The last ten years of this consumer-centered period (1960-1970) was, in reality, its mature stage. In that decade, only the computer had yet to reach its zenith, while all other innovative activities had already gone through development and entered into the style and modeling phase. For instance, without the constraints imposed by EPA environmental regulations, what remained to be done to automobiles beyond cosmetic changes in styles and models and an increase in horsepower to capture the market?

In this style and model change phase of product development, technology stagnated, and it became very easy for foreign nations to catch up, with or without organized technology transfer. Generally speaking, once a particular technological feat is demonstrated, it is less than half of the task for those in the general field of expertise to duplicate it. In fact, within an economic bloc (this bloc will probably be expanded to include the whole world very soon), it is highly desirable to let the developing nations learn new technology, because their lower labor cost can be enjoyed by the consumers of the forerunning nations during the initial catching up period. This advantage to the people of the forerunning nations is equivalent to charging tuition for the needed knowledge. Looking at the overall picture, the people in the catching up nations are consumers also and thus, according to the concept established by Henry Ford I, their added capacity for consumption would expand the total market turnover within the entire economic bloc.

In each major area of innovation, the catching up period may last 10-20 years, during which time the forerunning nation (usually the U.S. in the past) can enjoy the initial marketing and technological advantages; thereafter, competition becomes neck-and-neck, with the outcome often depending on where the people of each nation place their emphasis. For example, in the last decade, U.S. professionals were still indulging in "high technology," while Japan and West Germany, with no defense burden or aspirations in space, put their minds to consumer oriented industries and thereby showed significant results, which added to the trade deficits of the U.S.

It is interesting to speculate on how world economics would have evolved in the next 20 years and what role the U.S. would have played if the energy crisis had not become an issue. In electronics, following the calculator, the consumerized computer, and the adaptation of the micro-processor to automobiles and appliances, there could have been a flourishing market in sensors, which have yet to be mass produced as counterparts of the human nervous system (a laser may be considered a sensor); other sensors may include molecule detectors and pressure, acceleration, and temperature devices, etc.) It is interesting to note that as yet we do not have any sensor which comes close to the abilities of the human tongue or nose for detecting taste and odor, let alone equipment with the sensitivity of the bloodhound's nose, taste buds of sharks, or the sensory systems of insects in locating mates or food.

5. THE NEED FOR INNOVATION IN CONSUMER-CENTERED SOCIETY

Riding in the wake of one technological breakthrough after another within the span of 50 years, it is hard to believe that we are running out of major technological innovations to further enrich our lives; yet, there is nothing on the horizon and the checklist of our own biological, behavioral, and functional needs cited above seems to be quite complete. If this were the case, then we have to be content with innovative activities more closely oriented to users' taste (usually called "need"), instead of the more prestigious characteristic of scientific exploration marking the early stage of each major breakthrough.

To use one extreme example, when a new fabric is invented, there is a good deal of research into the molecular structure of the fiber to determine all the physical and chemical properties, so that the production plant may be developed and the fabric woven into beautifully designed material, such as chiffon. While this material may be prettier than the silk owned by queens of yesteryear, due to the uncertainty of consumer taste when it appears on the market, the manufacturer may be totally disappointed to discover that the fashion has changed to dungarees. On a larger scale, several billion dollars and numerous innovative schemes were employed by developers of the Bay Area Rapid Transit (BART) system, who later learned that the potential consumers were not quite ready to give up their private cars.

The market is, however, open to "novel" devices; for instance, Polaroid spent several hundred million dollars on the SX70 instant camera, which represented an improvement over their earlier models through the introduction of a new kind of film; it also featured an ingenious, folding, optical system and a motorized mechanism powered with a flat battery, which is packed inside with the film pack. When the first SX 70 had just gained consumer acceptance, Kodak, the giant in the industry, entered the market with its own version of an instant camera lacking a motorized mechanism but featuring the handle. This was potentially stiff

competition for Polaroid, in addition to their colossal, legal entangle-
ment with Kodak; however, as it turned out, both types of cameras are
selling well, and Polaroid was pleasantly surprised when 1976 turned
out to be their best year ever. Their primary difficulty was in increasing
the rate of production enough to meet the market demand during the
Christmas rush of 1976.

In this consumer-centered market, market demand hinges entirely on
acceptance by the public. The term "market need" is really a misno-
mer, introduced by the famed economist Adam Smith 200 years ago. At
that time, people were still struggling for subsistence, and, if a man
needed a plow and a woman a needle, there were just no two ways
about it. But with the onset of the consumer-centered society, the concept of need changed, and emphasis was placed on consumer acceptance
or appetite; for instance, a person may think he needs a big car even
though he may get by comfortably with a small one. After a while, he
may become addicted to big cars, and, when it seems that every one
else has a big car, he feels he must have one as if it were his birthright
until something new and "better" appears on the market to change his
"need." This "something better" usually consists of an improvement
in performance, which is distinctively indentifiable. Many products
were indeed introduced with the required "uniqueness" distinction
such as Polaroid film, the Xerox machine, and the IBM "Selectric"
typewriter (with which secretaries became enamored, despite the fact
that it is more than double the cost of conventional models).

6. THE MANY LEVELS OF INNOVATION

In this consumer-centered period, there were a few major technological
breakthroughs and numerous, significant technological innovations.
These were supported by myriad, innovative adaptations which were
then followed by skillful design, model variations, and style changes. A
major breakthrough, such as airplanes or computers, fills a void for
supplementing a particular, human biological function and represents a
giant leap in mankind's capabilities. In each case, during the introduc-
tory period, it was this prospect of making the giant leap that stimulated
a massive development program and induced investment. So long as
progress was being made, notch by notch, with the prospect of a giant
leap, there were eager investors, some of whom became losers and
some for whom the gamble paid off. The odds are not so important as
confidence in the payoff if one wins.

For the academic world, the early phase of each giant leap in technology represented a fertile area for research. Funds provided by private investors or government grants were easy to come by and had relatively few strings attached, and, since the path toward achieving the goal was vague, researchers were free to choose the way. Because a great deal of the effort was still in identifying the parameters of the natural (and unpatentable) laws involved, few confidentiality questions inhibited publication.

As each major, technological breakthrough finally matured and the promised giant leap became reality, the inducement for investment wore off. This was probably what happened at the close of the consumer-centered period which peaked around 1968 with the maturing of all other major, technological breakthroughs, except the computer. Maybe the next major technological breakthrough will appear on the horizon soon; perhaps it will be the use of nuclear fusion to supply abundant, inexpensive power though, at present, neither fusion nor any other form of new energy is quite ready for the all out push.

While only a major, technological breakthrough can set the investment mood into high gear, significant, technological innovations and innovative adaptations could and should be relied on for activating the market. The competition between Kodak and Polaroid in the field of instant photography illustrates that rivalry in business can be beneficial all around, because, in the consumer-centered society, market size is limited not by the physical needs of the consumer but by the consumer's psychological appetite. For the most part, this appetite is whetted by innovative schemes that the manufacturer injects into the product and, coupled to a degree, by exposure gained through advertising. Conversely, it is also true that competition is the best mechanism for generating innovation, which includes the coordination of design and style, depending on product type.

When in a consumer-centered society there is no energy crisis (but, to a lesser degree, other resource constraints*), the size of the commercial market is determined by only the appetite of the consumer. The commercial market, in turn, determines the job market after adjustments have been made for foreign trade, and foreign trade is determined by the relative strength in technological innovation, innovative adaptation, design, styling, and marketing of the domestic industry relative to its foreign counterpart. Such a view prompts one to observe that even though

* It should be remembered that energy can generate resources to offset depletion.

it is to the credit of the U.S. to have given the world most major, technological breakthroughs, after that feat was accomplished, there didn't seem to be enough technological innovation and innovative adaptation in the U.S. to continuously provide the consumer with attractive, new products in order to maintain the lead in the economy. To some degree, the U.S. fell behind Germany and Japan in this respect; one reason for this may be that it takes a greater diversification of systems, training, and talent to tackle consumer oriented, technological innovations and innovative adaptations than to develop major, technological breakthroughs for which the U.S. engineering community is famous.

7. INNOVATIONS NEEDED UNDER THE CONDITIONS OF ENERGY CONSTRAINTS

The energy embargo imposed by Arabs in 1973 produced a still more complicated and pressing situation. The overriding concern which we are facing now is energy conservation. While there are different views in Congress of how conservation should be carried out and at what rate, the outcome of an energy conservation policy, such as the one proposed by the Carter administration, appears to be certain. A prudent energy policy must gauge the rate of consumption according to our possible share of known fuel reserves over the remaining time span before viable, alternative energy sources are found. In any event, that share of fuel will not suffice, if the U.S. maintains the unrestrained, consumer-centered economic practices of the past 50 years. Before the energy crisis triggered by the Arabs, the unrestrained, consumer oriented economy seemed to have already run out of steam, due to the decline in innovation to stimulate the market, a fact which could be considered a blessing; otherwise, the demand for energy would certainly be even higher. On the other hand, the lack of market stimulation has already caused the above normal rates of inflation and unemployment, and any further cutback in the amount of energy demanded or in energy sources used seems to cause further increases in unemployment. For instance, if Detroit must switch from producing big cars to producing small ones in order to save energy, and if the labor needed for producing small cars is less, then simple arithemtic will tell us that there will be layoffs to match the energy conserved. Likewise, there will be a similar reduction in profits, which is of particular concern to the industrialist. The argument follows that with a reduction in profits, there will be a greater reduction in investment, causing a further contraction of the market and resulting in a snowballing, downhill trend in the economy.

Indeed, this kind of chain reaction is quite possible, especially if an energy crunch of severe magnitude occurs before the economic cycle has had sufficient time for cushioning in order to make the adjustment. In fact, what President Carter is attempting to introduce now are mild, man-made constraints to induce adjustment while resources are still quite plentiful and manageable. However, some people think that President Carter is overreacting and that his proposal might excessively burden a certain sector of the population, while many conservatives feel that government regulations inhibit innovation and therefore deter economic growth. Thus the question arises as to whether innovation is still a feasible method for stimulating the market under the mild and artificial energy constraints proposed by the administration.

8. CASE STUDIES OF INNOVATIONS WHICH DISREGARD ENERGY CONSUMPTION

To examine this question, it is interesting to review a few examples of product innovation in the past. Twenty years ago, a high fidelity, sound system was just on the horizon, and the frequency response of the loudspeaker was the weakest link. One of the problems was that the natural frequency of the speaker tended to be higher than the bass frequency desired. This is particularly true for a speaker with a smaller enclosure which is sealed to avoid interference; in this configuration, the air inside the enclosure behaved like a spring, thereby further boosting the natural frequency. One scheme experimented with added a negative spring to the speaker to reduce the stiffness of the system and lower the natural frequency. At the same time, Acoustic Research, Inc. took a different approach by loading the cone of the speaker with dead weight in order to reduce the natural frequency. This was indeed a very simple solution, even though the energy efficiency was poor because of the added dead mass—but energy was cheap, and the system became a standard scheme which is still in use.

A modern house was designed by a well known architect some 20 years ago in a Boston suburb. The owner requested thermal pane windows to conserve energy, but the architect dissuaded him from using them, because, according to his estimate, the cost saving on fuel was not enough to balance the interest on the additional expense incurred for their installation. Specifically, for aesthetic reasons, the dimensions of the windows designed by the architect were all nonstandard, which meant that the double glazed windows would be prohibitively expensive.

Some 12 years ago a major automotive company awarded MIT a relatively unconstrained grant of 1 million dollars to study automobile operations. At the same time at MIT, a small vehicle with an elaborate, active roll mode stabilization system was developed by a faculty member with his own support, who thought that cars were getting too big. This system and the study were naturally linked and might have yielded tangible results, but the automobile company, while accepting all paper-study types of proposals, was not interested in small vehicle development. This negative attitude was not based on a concern for possible failure but rather on possible success at a time when they wanted to cultivate the big car market.

These examples illustrate that energy-inefficient products are frequently developed out of expedience. It is conceivable, however, that within certain limits innovation can be applied to modify products to achieve greater energy efficiency, while maintaining the level of marketability. Needless to say, while cheap energy offers more options in the design of any product, to maintain product marketability in an energy constraint situation, the best countermeasure is still to depend on innovation.

9. THE STRUCTURE OF A MODERN SOCIETY BASED UPON THE TIME BUDGET OF ITS MEMBERS

The interplay between the well-being of a society vis-à-vis innovation, the energy supply, or, for that matter, all environmental factors, including resources and waste disposal, can be appreciated by examining Figure 1.1, which shows the dominating factors in an economic system such as that in the United States. The center of this diagram shows primary human roles, namely those of consumer and worker which are linked by earning and spending. In a primitive society each family is self-contained, in that earning and consuming are often confined within each family; in a more complicated society, working and consuming are interwoven in a very elaborate fabric, though the essence of these two primary functions remains unchanged.

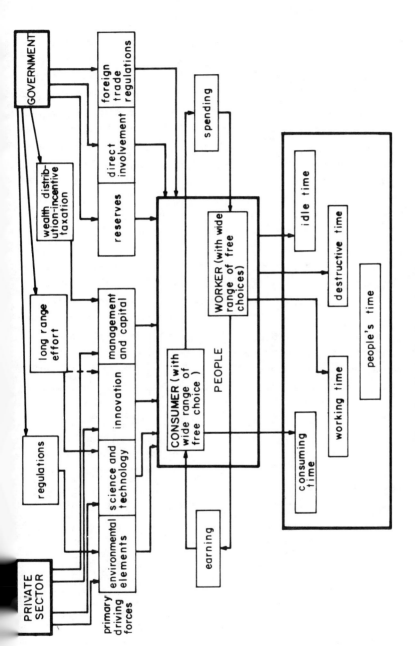

Figure 1.1. Dominating factors in modern society.

At the bottom of the diagram, "people's time" is recognized as the sum of four major blocks of time which are utilized by all people in a particular society: namely, "consuming time," "working time," "idle time," and "destructive time." Consuming time represents the time each individual spends enjoying the fruit of his labors (shopping, eating, traveling, studying, etc.) and usually involves expenditures, in direct contrast to working time when money is earned; idle time involves neither earning nor expenditure; the last item, destructive time, represents the time utilized by certain groups of people to generate social disorder and the time consumed by society to counteract that disorder.

The well-being of a society may be described in terms of the distribution of time among the four major activities listed in Figure 1.1. It is quite obvious that the well-being of a society would markedly decrease if idle time (such as unemployment) and destructive time were increased; thus, there is a time budget for the society which could be even more important than a monetary budget, such as the GNP. Conceptually, "consuming" and "working time" are balanced to achive equilibrium. People need time for consumption and relaxation, which is made possible by the effort exerted during working time in association with the introduction of four prime driving forces shown in the diagram and identified as: Environmental Elements, Science and Technology, Innovation, and Management and Capital. Normally these driving forces are provided by the private sector, which is represented by the leaders in various fields.

Over the last century, the quality of life in the U.S. has undergone significant improvement, and a wide variety of conveniences to satisfy people's needs were developed with an immensely wide range of options to suit individual, consumer taste. It is interesting to note that some appreciate this range of free choice in either consuming or working* more than the absolute quality of the items involved. However, during the same period, great progress was made in upgrading working conditions by eliminating the 72-hour week and establishing the 35-40-hour work week (including two daily coffee breaks) in air conditioned shops, with numerous holidays and vacations. Automation and

---

* This desire for a "range of free choice" varies from nation to nation and depends very much on traditions and social environment. The use of "range of free choice" as a measure of the quality of well-being is typical of thinking in the U.S.

unlimited energy resources made it possible to vastly increase the supply and variety of consumer products, while reducing working hours by one half; and household chores were simplified, freeing more time for consumption and relaxation.

The shift from a "working time" dominated society to a "consuming time" dominated society was a slow process, involving many fine adjustments from the introduction of automation, which is generally considered a sign of progress, to the featherbedding tactics used by some labor organizations, which are often regarded as setbacks. However, both were essential in maintaining the balance between the two principal blocks of people's time during the transition period. It is interesting to note that while automation made it possible for the consumer to enjoy vast amounts of wealth, it was the continuous improvement of the product which, by arousing the interest of the consumer, justified automation. To those responsible for production, automation means efficient use of working time and is, therefore, a good thing. But if the market is already saturated with a particular product, then automation causes layoffs and subsequently an imbalance in the time budget of the society. An increase of idle time which breeds destructive time then occurs. Thus, innovation must be used continuously to adjust the balance of the two major blocks of time to prevent the occurrence of idle time and consequently the increase in destructive time. In the U.S., the adjustment of the balance of the time budget has not been done by means of a master plan but rather has been primarily by the vision and motivation of a large number of individual entrepreneurs.

Generally speaking, automation is introduced when a particular product enjoys wide customer acceptance, with a surge in demand; otherwise this is a destabilizing influence on the balance of the two blocks of people's time (consuming and working). The destabilizing influence of automation did not occur in the past, because innovators were able to introduce new products continuously, thereby maintaining the balance.

The "consuming time dominated society" is a relatively new phenomenon, probably unprecedented in all human history. Most people within the society, including leaders in various fields, still consider work a major virtue and make plans according to the needs associated with that virtue, which, in turn, generally means more efficiency in task performance. In reality, however, it is the balance between the two blocks of time that is most important, and, as the amount of consuming time increases, it becomes more and more the dominant or problematic factor.

An essay entitled "The Big Puzzle: Who Makes What and Why," (*Time*, June 13, 1977), sheds some light on the expansion of consuming time in U.S. society. In this article, *Time* cited the vast difference in pay among many professionals and lamented the fact that "many Americans are displeased by what certain disparities seem to disclose about social values. For these prevalent pay differentials, taken as an index of the social soul, seem to prove, for example, that the nation cherishes professional teachers far less than professional athletes. Or, more broadly, that society generally values members who do its most serious work not nearly as much as the actors, clowns, and jocks whose task it is merely to distract and amuse." A few examples of the prevalent pay differential cited by *Time* include:

Trial Lawyer, Washington, D. C.	$500,000/year
Marlon Brando	$2.25 million/12 days
History Professor at	
University of Virginia	$30,000/year
Johnny Carson	$3 million/ year
Mohammad Ali	$5 million/night
Minister, United Church	
of Christ, Chicago	$10,500/year
Violinist, Boston Symphony	
Orchestra	$24,800/year
Bus driver, Atlanta	$13,500/year
Chairman, Dow Chemical	$453,000/year
Basketball star, New York City	$325,000/year

One logical conclusion reached in *Time*'s article is ". . . this handy method of social soul searching is not reliable. Far more directly, income differences reflect the operations of the marketplace." . . . .

After considering the nature of the "consuming time dominated society," it becomes easier to understand the existence of such a marketplace, where a large block of people's consuming time is to be used up. The amount of compensation received by athletes, actors, or any other entertainer simply reflects how much of the people's consuming time is used up by their services. Television broadened the market of entertainers in the same way that automation and mass production methods broadened the market for industrial products. Of particular interest in *Time*'s table of earnings for various professions is the very high pay received by a certain trial lawyer. This situation seems to violate the rule illustrated above, since a trial lawyer produces neither goods nor draws large numbers of spectators.

The existence of the legal profession reflects the emphasis on "free choice" by people in consumption and work as a way of life in American society. "Free Choice" usually causes conflicts or collisions between parties aiming at the same goal, which requires intercession of the law and hence creates the need for the legal profession. For instance, when no-fault insurance was introduced in Massachusetts, the cost of premiums dropped significantly, with the legal profession as the principal loser. In general, free choice is a costly system, not just in terms of legal costs but in a very broad sense. It is expensive, for example, to have annual, model changes in cars, a luxury to which the U.S. is accustomed and which it can still afford.

Over the past century, the U.S. economy enjoyed unprecedented prosperity, with the result that working time shifted gradually toward consuming time, while the range of free choice progressively increased to a level previously undreamed of. Science and technology and innovation were two primary, driving forces behind this movement with "accumulated capital" providing the tools and muscle. In the diagram discussed earlier, these three elements were shown together with the fourth factor, the unrestrained flow of environmental elements, (resource supply and waste disposal) as the driving force. Up to the end of the consumer-centered period (1920-1970), environmental elements in the U.S. played the role of a noninterfering benefactor.

Among the four primary, motivating forces shown in the diagram, "science and technology" represent accumulated knowledge, while innovation respresents fresh, new ideas. In the language of mathematics, science and technology are the integration over time of man's innovation within the various disciplines, while, inversely, innovation is the derivative of man's accumulated knowledge. As a rule, it is the integrated effect which has significant impact, but it is the derivative which yields fast response to the changing environment. Furthermore, innovation is not limited to progress in science and technology; rather, it generally represents all fresh approaches to every human endeavor. In the period of increasing consuming time in U.S. society, it was innovation which provided myriad kinds of products to attract the interest of the consumer.

10. NEED FOR INNOVATION IN AN ENERGY-CONSTRAINING AND TIME-BUDGETING SOCIETY

As described earlier in this chapter, innovation in the U.S. was responsible for numerous, technological breakthroughs in the past; in the future, more innovation will be needed to herald the breakthrough in nuclear fusion (or other energy sources) to give us unlimited power. At the same time, less sophisticated innovation is also critically needed to supply the consumer with products which are equally attractive, or more attractive, than those currently available but consume less energy. Indeed, all three examples cited earlier—the loudspeaker, the automobile, and the window—can now be designed to conserve energy while maintaining their performance levels.

For the loudspeaker system, there are now much stronger permanent magnets than were previously available which reduce the consumption of current. Furthermore, a pulse width modulated amplifier can be efficient as well as low in impedance, thereby cutting down the internal power dissipation so that the amplifier can virtually ignore the mass loading effect of the speaker as a source of power drain. As for the automobile, the law against the "gas guzzler" elicited some complaints from the industry, but, painful as it might have seemed at first, gas mileage of the average car has improved remarkably over the past few years, and the trend seems to be gaining momentum. Weight reduction was accomplished, in part, by substituting aluminum and plastic for steel, and engine efficiency was improved through the introduction of sensors and microprocessors. In the case of the house which did not receive Thermopane glass 20 years ago, a solution was found through innovation. The owner was forced to invent a new way of attaching a thin, transparent film to the windows which could be removed and easily stored in summer and, in winter months, provided a form of trapped air insulation. His architect friends were also impressed with the aesthetically pleasing appearance of the film and even suggested that the new scheme might open a new market.

11. THE ROLE OF GOVERNMENT IN THE DEVELOPMENT OF INNOVATION

In Figure 1.1, the role of government is assigned to the upper right hand corner. Conceptually, in any democratic system, the role of government is to serve by supporting citizens as they do society's work. This supporting role may be characterized by: 1) judiciary and incentive functions pertaining to interrelations between people and groups of people; 2) managing public works which are too *large* to be handled by individuals or groups of individuals; 3) managing works pertaining to events in the distant future, beyond the scope of any individual or group of individuals.

While it is not the purpose of this chapter to investigate the principle of government, it is pertinent to note how innovation was treated by government policy vis-à-vis its counterpart science and technology. Traditionally, science and technology have been viewed as a manifestation of the long-range progress of a society and, for this reason, government agencies, including the National Science Foundation, have taken a strong interest in their development as a social driving force. Through these agencies, the bulk of the work is diverted to, and conducted in, various universities and institutions.

Innovation, on the other hand, is a fragmented activity. Those innovations pertaining to science are logically included with science and technology, but those innovative activities more closely related to product development have, logically, always been treated as part of industrial operations. The interesting twist is, however, that even though innovation in itself is not a long-term proposition, as are science and technology, the training of innovators in a certain area requires many years of development.

In this "consuming time" dominated society, we desperately need many innovators who can generate a large quantity of salable products which burn less fuel while opening up more jobs. We are already behind Japan and West Germany in the training of such innovators, largely because of the emphasis placed on high technology in the space era. The question is, therefore, whether or not training innovators is a task which justifies national attention (represented by the single dotted line in Figure 1.1).

12. HOW THE NEED FOR INNOVATION FITS INTO VARIOUS ASPECTS OF INDUSTRY

In broadly claiming that innovation is critically needed in the U.S., one might encounter quick rebuttal from the electronics industry, where the "boom, zoom, and bang" phenomenon described by Seymon Schweber in his open letter to the general public (*Electronic News*, June 13, 1977) is evident. Schweber feels that there is too much innovation going on in the electronics field instead of too little. In his letter, he cited the rapid rise and fall of companies dealing with handheld calculators, digital watches, C. B. transceivers, and T. V. games and warned those who are just beginning to climb on the micro-processor bandwagon. The warning is most timely, but is too much innovation to blame, or rather, should blame be attributed to lack of training in innovation, which prompts eager and ill prepared entrepreneurs to rush the floodgate of sudden product popularity?

There is no simple answer to these questions, because those who were able to ride with the tide of a booming market and zoom to the top in the initial phase of fast growth are indeed capable innovators. This recent phenomemon in consumer electronics is really a repetition of the pioneer days of the aviation and auto age: Each of these technological breakthroughs was followed by a flood of emerging industrial firms; for example, at one time, there were several dozen, small automobile and airplane companies. Many faced the boom zoom bang phenomenon described by Schweber. These companies swamped the market like the throng of runners at the starting point of the Boston Marathon but, as in the Marathon, only a handful reached the finish line. This high elimination rate is, however, a natural process of innovation. On a smaller scale, each individual in searching for a marketable product must go through a selection and discard process to narrow his choice among many possible ideas, test models, and design revisions; for the entire market, the boom and bust phenomenon is only the continuation of the selective process.

The famed Nobel laureate, Dr. Linus Carl Pauling, was once asked by an admirer how he could be so productive in generating ideas; he answered that it was not because he could generate good ideas but rather that he had many, many not so good ideas and a talent for selecting and pushing only the better ones. Thus, while the boom and bust phenomenon may be painful for the losers, it is nevertheless a healthy and necessary process for achieving the final product which benefits society.

The question relating to training in innovation is even more elusive. First, there is controversy over whether innovation should be trained, let alone whether it can be. For example, if training in innovation is intended to make the innovator cautious and refrain from rushing the floodgate, even though there may be fewer casualties on the market battleground, the end result certainly does no service to society. On the other hand, if training is meant to make the individual innovator more prolific in generating ideas, as well as more skillful in the selection process, he would occupy the pole position and enjoy a better chance of winning the race. If the majority of innovators are prolific in generating ideas, as well as more skillful in the selection process,there may be even more lively competition, but there will be a general elevation of the social benefit.

13. THE USE OF "PARAMETER ANALYSIS" AS A METHODOLOGY FOR TRAINING IN TECHNOLOGICAL INNOVATION

All creative minds probably go through similar mental processes in producing a masterpiece. A composer creates his music by deliberating on the musical notes, the structure of the chords, and the tonal quality of the various instruments. As an example, a composer inspired by some external stimulus, be it the gentle rustling of a breeze or a storm of great turbulence, may mentally "hear" music. If there happens to be a musical instrument around, he may try out a few notes, then modify and expand them into bars and passages enriched with harmony and chord, maybe even borrowing from other existing melodies and adapting the style of certain classical or folk music traditions until, perhaps, a symphony is born. Whether or not he creates an acclaimed new work, the process of creation involved the juxtaposition of certain musical building blocks which were stored in his mind. The creative process also involves selection and testing similar to the invention process described in the last section.

Any creative person, whether a writer, scientist, inventor, or entrepreneur, must perform creative tasks by selecting certain mental building blocks after many others have been discarded. Acquiring these mental building blocks, skill in selecting them for use at the right moment, and developing a critical eye for rejecting the ill-fitting ones, lie at the root of creative power. In short, let this process be identified as parameter analysis methodology.

It is interesting to note that creative writing, music composition, dance, and athletics are taught in schools, even though achievement in each activity hinges heavily on individual talent. By contrast, in the conventional curriculum of engineering and management schools, very few courses emphasize developing creativity in invention and entrepreneurship. Likewise, in industry creativity in invention and leadership in entrepreneurship are uncertain occurences. The need for a better environment in which to cultivate these talents was expressed forcefully in a letter written by Dr. Tom Butler, Director of Research and Vice President of AMF, Inc.

"..... My general observation regarding our attempts at joint efforts is that we have difficulty in reducing a new concept to practice. Our business unit engineering staffs are directed primarily at relatively small improvements from one model year to the next. The pressures of this activity make it difficult for us to give sufficient attention to significant new developments. In order for a new development to be absorbed quickly into our business unit, it appears that the development should be so complete that all that remains to be done is some production engineering. Perhaps we should devote more attention to this problem area. For example, who should guide the development of the concept and how should it be financed until AMF is ready to absorb it into a business unit. Perhaps more could be done in this area which also would provide practical experience to business and engineering students . . . ."

This letter was written in connection with the development of a sports product which has some distinctive properties in comparison to existing equipment. A laboratory model of the new device was developed at the M.I.T. Innovation Center, and a license agreement was signed with A.M.F., but the next step—development of an engineering prototype—ran into a series of difficulties. As expressed in Butler's letter,- the critical issue is not the development, or lack of development, of a particular product, but the need for a system within the company which inspires a level of innovative accomplishment higher than that required for regular model change. Such a need is certainly not unique in A.M.F.; it is, in fact, fairly representative of the industry; however, Dr. Butler is to get credit for identifying it clearly.

Sports equipment is after all technically uncomplicated and "need" is easily specified; furthermore, sporting goods are end products for the consumer. If a barrier can arise in the development of an innovative sports product, how much more complicated would the process become for a more sophisticated system, such as an anti-skid brake for automobiles and trucks?

Along these lines, Dr. Lamont Elting, Director of Research for the Eaton Corporation, made the following comment:

"..... Your observation that you "can only get involved in consumer-type products" is interesting. That may be due to ease of defining the boundary of the product and the contribution made by the individual inventor, as you indicate. It may also be because of the available specific and extensive definition of the need for the potential inventor. Certainly, need must be defined and well understood for invention to take place. Perhaps attention to such definition of need will stimulate invention of larger systems. It's ironic that participation in a field seems required for ready familiarity with the needs, yet it produces some "intellectual antibody" responsible for many inventions coming from outside the industry. . . ."

Eaton has the reputation of being an innovative company: Aside from its anti-skid brake and leadership in the controversial, air bag safety system, Eaton has recently displayed eagerness to get into microprocessor technology, all of which represent a positive, innovative climate. Yet Dr. Elting could not help but observe the "intellectual antibody" phenomenon which deters advancement in technological innovation.

The exercise conducted by Y.T. Li and Bernard Gordon dealing with computerized axial tomography (described in Chapter 3) demonstrated that some form of parameter analysis was used by the established entrepreneur to innovate and the methodological exercising of this skill may even be useful to the most skillful and gifted innovator in his own field of expertise. Since computerized axial tomography is at least on a par with, if not more sophisticated than, the anti-skid brake system, one may wonder why parameter analysis methodology would not be equally effective for an anti-skid brake system or its equivalent, thereby avoiding the "intellectual antibody" described by Dr. Elting. Thus, it is the objective of the remainder of this work to examine how the parameter analysis concept can be implemented in the existing university environment and industrial operation, as well as to seek a framework for promoting a mutually beneficial collaboration between these two groups.

2
PARAMETER ANALYSIS AS A METHODOLOGY FOR TRAINING TECHNOLOGICAL INNOVATORS

Generally speaking, all new, creative works are evolved from elements of prior art: A composer uses the conventional musical notes and the tonal quality of ordinary instruments as his building blocks; a writer can relish words, phrases, and philosophical thoughts. For technological innovators, the creative interludes are probably much broader than in either of the two professions mentioned above; Figure 2.1 may be more effective than simple words in portraying their involvement.

1. THE SPECTRUM AND STRUCTURE OF ACTIVITIES IN TECHNOLOGICAL INNOVATION

In Figure 2.1, technological innovation is incorporated into the shape of a cornucopia—the traditional symbol of abundance. At its tip, the system is fed by the body of scientific knowledge from which technological disciplines have evolved; down the line, there are industrial operations which exist to bridge technology and the fulfillment of social needs; and, finally, products materialize to complete the process. The dotted line linking the five stations identified by the Roman numerals encompasses the various stages of technological innovation and symbolizes the fact that, for each product placed on the market, germination must occur at station I with the recognition of a social need; germination is followed by the perception of possible, innovative solutions at station II; station III represents an iterative synthesis process, where the various technologies are blended to create a new product configuration which will then be tested successively to eliminate all questionable side effects and verify the novel claims. If all goes well, the innovative idea will gain further financial support and will be brought to fruition through the development phase (station IV) and the general industrial operation (station V).

Thus, the critical juncture in technological innovation is the three-way intersection at station III, where technological information meets the perception of a social need in order to initiate the downstream push toward commercialization. The most frequent complaints regarding unsuccessful, innovation processes illuminate the importance of this three-way intersection; for example:

- Many industrial giants lack the ability to innovate:

  Their operation places too much emphasis on short term profits, characterized by operations at station V, which in many respects is incompatible with the upstream initiation of new products.

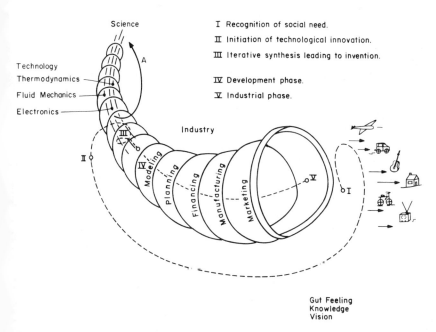

Figure 2.1. Dominating parameters and functional relationships of science, technology and innovation.

- Many small inventors cannot market their innovations:

  They have no concept of the downstream operations from station III to station V; their inventions aren't realistic; they don't know how to interface with the downstream proponents.

- Despite all the grandiose proclamations of technology transfer, many government agency and university owned inventions are idling:

  These inventions are usually fallout from scientific research. Although technically sound, they have little connection, with any need, nor has there been any consideration of the downstream problems. Thus, each may require another major innovation to be matched to an appropriate social need, or each may have to be modified and reinvented to become practical.

2. THE CONVENTIONAL ENGINEERING EDUCATION PROCESS

The cornucopia diagram also puts technology in its proper perspective. The dominant characteristics of technology are illustrated by the neatly drawn parallel lines, which represent the distinctive nature of each discipline in engineering science which bloomed profusely in the last 50 years. The solid line A marks the feedback from the junctional station III of Figure 2.1 to the fountainhead of knowledge, where science represents the primary goal of education in engineering science, and illustrates the regenerative process in the development and enrichment of various technological disciplines. Research methodology in engineering science focuses on parameter identification similar to that in basic science.

Great philosophers and scientists often spend long years of painful investigation in order to identify the key parameters within a large collection of other more obvious, natural phenomena. For example, Newton's first law of mechanics, $f = ma$, describes the three basic, physical parameters of force, mass, and acceleration and their relationship in a simple and elegant manner. These parameters exist in our daily environment, along with innumerable other parameters, but how many of us can make the recognition before it is pointed out? Indeed, outstanding students in science and engineering not only strive to assimilate the knowledge of the masters, but, aided by myriads of modern instruments and apparatus, attempt to apply the methodology of the masters to open up new frontiers. Great philosophers and scientists are also great teachers. Their findings, in the form of well-defined parameters and the relationships between these parameters, are generally applicable to similar situations which repeat themselves over and over again; thus the classic book *The Principia* is as valid a reference text now as it was several hundred years ago when it was written by Newton.

In this same spirit, most of serious research work in engineering science reveals a knowledge of the regenerative process, and most of the esteemed textbooks in engineering science are organized to generate knowledge efficiently in order to help future researchers. When this regenerative system is in full swing, the most acceptable means of demonstrating one's creativity and thereby gaining recognition and esteem in academia is through publication of research findings. Well developed technology disciplines make it easier to improve the performance of technological innovations even though innovation is not, at present, a regular part of the university curriculum.

Figure 2.2 illustrates conceptually the various ingredients in engineering education and to a large extent the present mode of operation of the technology community. To the left of the diagram, natural phenomena are studied and identified through research in engineering science, according to parameters and parameter relationships, and are then organized into various technology disciplines, fields, areas, etc. Near the bottom of the diagram, a roll of circles represents the various system configurations which some innovators devised to meet social needs by satisfying a certain set of product specifications shown at the bottom of the diagram.

In the conventional sense, innovation is the result of a special talent and is therefore shown in Figure 2.2 as introduced from outside the information flow stream. If a particular innovation is indeed outstanding, the resulting system configuration may readily satisfy the specification of the need without interfacing with technology. However, most innovative devices have their performance improved through the use of engineering design, and an experienced designer, as a rule, innovates during the design process. In its primary sense, a design starts with an existing configuration which meets a given set of specifications, and it is the designer's function to alter the value of the parameters within this configuration in order to modify or improve performance specifications. An example of design modification was enlarging

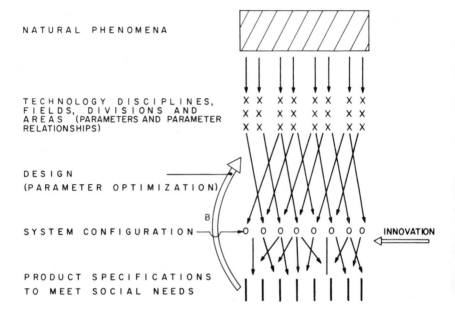

NATURAL PHENOMENA

TECHNOLOGY DISCIPLINES,
FIELDS, DIVISIONS AND
AREAS (PARAMETERS AND PARAMETER
RELATIONSHIPS)

DESIGN
(PARAMETER OPTIMIZATION)

SYSTEM CONFIGURATION   INNOVATION

PRODUCT SPECIFICATIONS
TO MEET SOCIAL NEEDS

Figure 2.2. Elements in conventional engineering education.

the DC-10 airplane to a "stretched" DC-10 to accommodate more passengers. Another design objective is to modify the parameters of an existing model to accommodate improved components or advanced material; a typical example is the introduction of composite material, such as graphite fiber and epoxy, to airplane construction.

Design function is represented by the wide arrow pointing upward from the bottom of the diagram in Figure 2.2. In a typical design operation, the solution is usually the result of some engineering trade-off. For instance, a "stretched" airplane may be more economical than the original because of its ability to accommodate additional passenger seats but, on the other hand, it sustains extra weight; accordingly, there must be an optimum length which will yield the highest cost-effectiveness. This cost-effectiveness factor is often known as the performance function, which is defined as the sum of all desirable features, each being properly weighed to establish the "goodness" of the system, which is then divided by the total cost. Thus, the objective of the design is to search for a set of configuration parameters which will optimize (maximize the performance or minimize the cost, etc.) the performance function, and the procedure is therefore called parameter optimization, as noted in Figure 2.2.

Conceptually, all engineering systems may be optimized through mathematical manipulation to arrive at the exact optimum solution, if it exists. In practice, however, design optimization is often done empirically to reach for the near optimum. This is in fact another optimization process, because the labor saved in taking the shortcut is probably more than balanced by the sacrifice in performance as a result of being slightly off the exact optimum. The fact that an exact optimum does exist makes engineering design a professional function, which means that the solution can be obtained by following a certain set of rules which can be documented and transferred. Thus, in conventional engineering education, the study of the basic parameters and parameter relationships in various technology disciplines provides the foundation of knowledge. Advanced study in engineering then covers research to provide the regenerative function to enrich technology itself, with professional engineering as an alternative route to specialization in design, with the goal of improving somebody else's innovation.

3. THE CONVENTIONAL MODE OF TEACHING IN MANAGEMENT SCHOOLS

Like engineering, management is divided into three basic phases: The first phase instructs the student in the tools necessary to become a team member in a business operation; tools needed in management include accounting, banking, contracting, financing, production management, etc. Conceptually, these subjects deal with parameters and parameter relationships as in the various technology disciplines. The second phase in teaching management provides the student with experience in decision-making, which is analogous to aiming at an optimum solution in engineering design. Mathematical modeling of parameter optimization as applied to business is known as operations research. A more practical approach, which is equivalent to an "engineering trade-off" aiming at "near optimum," is often conducted through the "Case Study" approach pioneered by the Harvard Business School. The third phase in management training uses statistical studies of certain groups of business operations to identify the key parameters of the individual groups; in essence, this phase of training serves the same function as research work in engineering science. Recently, quite a few excellent research papers were presented identifying certain characteristics of industrial innovation. However, none of these are intended to train technological or business innovators except as researchers.

In the last 60 years, or during the consumer-centered period, as identified in Chapter I, the United States saw unprecedented activities in business innovation parallel the rapid growth in technological innovation. Henry Ford I's policy of paying high wages to his workers in order to induce them to become consumers of his product was a business innovation. "Buying on credit," which started after World War II and gave the U.S. economy a sudden boost, was another business innovation. Supermarkets, credit cards, fast food, and franchised stores and hotels were all business innovations. In the case of Henry Ford's famed Model A and Model T, success was a total entrepreneurial experience, including supreme accomplishment in both technological (even though there was no significant invention in either case) and business innovation. But, by and large, most business innovations are introduced without particular technological innovation. Likewise, many successful technological innovations make no claim on business innovation, even though the loop around stations I to V of the cornucopia diagram (Figure 2.1) relies heavily upon business operation. In the following chapter, the role of professional management in technological innovation will be elaborated further.

4. PARAMETER ANALYSIS AS A METHODOLOGY IN TECHNOLOGICAL INNOVATION

At the end of Chapter 1 (and in the last sections of Chapter 3) the use of parameter analysis as a methodology for innovation was briefly introduced. In a broad sense, it begins with recognition of a social need and results in a search for a technically feasible solution. Now, after a survey of the conventional educational process, it is possible to put the parameter analysis methodology into better perspective.

Parameter analysis has one unique objective—to come up with a new configuration for a marketable product. This specific function is an important feature which is absent in conventional engineering education (shown in Figure 2.2), as well as in conventional management education, discussed earlier. Figure 2.3 illustrates how this missing feature may be added to conventional engineering education. Here, a dotted circle, superimposed over an existing circle, represents the attempt to generate a new innovative configuration.

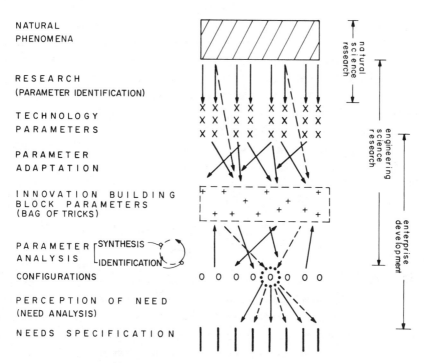

Figure 2.3. Elements in engineering education for enhancing technological innovation.

Also added to Figure 2.3 from Figure 2.2 is a group of crosses representing building block parameters. Building block parameters are introduced to synthesize new configurations and to relate to the existing technology parameters and natural phenomena through a series of crisscross arrows labeled parameter adaptation. It is to be noted that most technological parameters were developed more for understanding natural phenomena than for generating new, man-made configurations. The reverse is true for building block parameters, which were assembled for convenience in generating new system configurations to satisfy new sets of product specifications. For example, "impedance matching" is a very useful conceptual parameter in the design of many man-made systems; it is also related to many nature phenomena involving energy transfer. However, in conventional teaching of electrical engineering, it is mentioned in circuit analysis at the end of a sequence of parameters, instead of being identified as a useful building block common to many man-made systems.

Most man-made systems can be improved over existing systems through the processes of mutation and progressive iteration. The mental process involved focuses on the study of similar types of existing systems and thereby identifies certain key parameters which are unique to the objectives of the new configuration whose specification is established by the need. This kind of parameter identification from existing, man-made systems represents another source of building block parameters. From these building block parameters, as well as from those adapted from technology parameters, appropriate parameters may be chosen to synthesize new product configurations. This iterative use of parameter identification and synthesization is herein called parameter analysis methodology.

Most technological innovation takes a long time to develop. Some of the pertinent building block parameters were identified specifically for the synthesis process which immediately follows parameter identification. At other times, useful building block parameters were drawn from the innovator's bag of tricks (mental collection of building block parameters) gathered through earlier observation of new devices. Developing the habit of identifying key parameters of outstanding, man-made systems or conventional technology disciplines is half the job of developing proficiency in parameter analysis or technological innovation. For instance, in the study of computerized axial tomography, pre-; sented in Chapter 3, the thinking process began with identifying the information contained in the single x-ray beam, as well as the key parameters of the conventional, longitudinal sectional x-ray machine. From these parameters, the conceptual parameter of commutation evolved, and a conceptual model for the CAT scanning system was synthesized.

In the last section of that study, the parameter analysis process was carried further with the identification of the partitionability of the two non-intercoupled signals originated by the orthogonality of information in the x-ray beams. Also identified was the statistical nature of the information. These last three conceptual parameters are more general than the parameters identified earlier and can therefore offer broader coverage in a patent application should that route be adopted. They certainly can provide more freedom in the synthesis process with the possibility of yielding a more cost-effective CAT scanner.

The layout of Figures 2.2 and 2.3, as well as the feedback loops of Figure 2.1, illustrate that engineering science has a built-in, self-regenerating feature, whose prime objective is focusing the teaching of technology on the training of future researchers. The structure of technology in the form of disciplines, divisions, and areas, as well as the development of all logic and exercise problems, are most effectively molded for this prupose. Even the objective of teaching design in the university seems to be providing interesting examples to illustrate the power of engineering science and is seldom intended to serve as a medium for teaching the process of generating new innovative configurations.

As for successful inventors and entrepreneurs themselves, they sense the parameters of each event as it confronts them. Each may grasp the building block parameters in his own intuitive manner and then relentlessly repeat the synthesization process until a satisfying solution is reached. For these individuals, the solution is the focal point of reward and satisfaction, and few are able or desire, to consciously retrace their train of thought once the solution is reached. But consciousness of the existence of a group of building block parameters as the possible media for innovation can be a great help: It frees one from deep-rooted training in engineering optimization and thereby prompts recognition of only those technological parameters associated with existing configurations.

Building block parameters have a much closer tie to product specifications than do technological parameters; for example, "parts count" can be treated as a building block parameter but definitively not a technological parameter. In one particular instance when an unconventional windmill was to be evaluated, some evaluators paid a great deal of attention to making parameter optimization of the rather significant, aerodynamic properties but commented only casually on the complicated nature of the system compared with the conventional type. However, as it turned out, the new system had ten times the parts count of the conventional model. It then became quite obvious that this particular parameter was the dominant issue rather than the aerodynamic effect; the moral is: A worthwhile innovation does not need the fine tuning of parameter optimization to show its advantages or disadvantages.

The parameter analysis approach is not limited to the initial phase of innovation or invention; it is applicable all the way from station I to station V of the cornucopia in Figure 2.1. The use of parameter analysis to generate a conceptual model which will enable an industrial leader to lead a technological innovation team will be discussed in the next chapter.

5. PARAMETER ANALYSIS APPLIED TO A SIMPLE PARTY GAME: AN EXERCISE

Parameter analysis is defined as a broad training methodology designed to cultivate the habit and ability to search for dominating parameters within new situations. These dominating parameters are not immediately apparent but have more influence on the outcome than those which are obvious; they also have generalized properties which are applicable to classes of situations, rather than to specific situations. If one is observant, he will find many interesting situations worthy of parameter analysis, and, to illustrate this point, let us use parameter analysis to tackle a simple party puzzle which may be glorified as game theory.

The puzzle is: Four cats, A, B, C, and D are placed at the four corners of a 10-foot square. Each cat is to chase the next, moving clockwise. If the cats all move at the rate of 10 feet per second, how long will it take for the cats to meet? Since this is a party game, the contestants must act fast, and those with quick minds may shout out the answer in a few seconds. The reader is welcome to pause here and give it a try. . . .

In a typical group, there will be some with considerable mathematical skill who will quickly see that the tracks adopted by the cats would be in the form of spirals. They will seek a method of determining the length of the path with the notion that the solution time is the quotient of the length of the spiral divided by speed. Unfortunately, this group will not be able to come up with an answer before the party is over. Presently, someone will call out the answer of 1 second; someone else will challenge with an alternative answer of 0.7 seconds; a third person will protest that the cats can never meet.

This is a good party game, because with its simple statement it provides intellectual challenge and a good deal of heated argument. The 0.7 second answer seems to be the shortest time possible, if the objective is to direct the four cats to collide. But the original statement asks each cat to chase the next cat in a clockwise direction. If the path is circular, then the third person must be right, because in this manner the cats would never meet; for a square, however, the first answer seems to be accurate. In reality, the first answer is correct, but few who arrive at this conclusion can explain how they did it, and why it is so.

First, we must assume that each of the four cats is doing exactly the same thing at all times. This would mean that their relationship to one another at any instant is the four corners of a shrinking square; this phenomenon may be called the *law of similarity*. (Pertinent parameters are underlined.) The cats are instructed to perform a chase. In doing so, they must have precise guidance instruction, and, if the cats are mechanized, there must be some control law. In astronautical maneuvers, such as a rendezvous between two space ships, one possible control law is *dead reckoning*, which simply means aiming one's direction of travel continually at the target, as if it were a stationary or "dead" object. Since all of the cats are moving orthogonally to one another because of the square pattern (note that this orthogonality is similar to the parameter mentioned in the last section in the discussion of Chapter 3), their converging speed is equal to the speed of each pursuer. This is because the velocity vector of the target cat is perpendicular to the pursuer's vector, so that the target's motion does not affect the relative pursuer-to-target progress. With an initial distance of 10 feet between cats and a converging speed of 10 feet per second, the second solution is obviously the correct one under the assumption of the dead reckoning control law.

With the control law identified as an important parameter, the game leads to further debate as to what the solution might be if a different control law is assumed. The debate can continue indefinitely. The moral of this example is that too often we make judgements intuitively, without adequately examining the pertinent parameters. One of the maxims when using parameter analysis as a training method is to habitually define and label the parameters whenever a fascinating situation arises.

6. A SIMPLE ILLUSTRATION OF THE EFFECTIVENESS OF PARAMETER ANALYSIS IN INNOVATION

One recent newsworthy event in technological innovation was the successful manpowered flight of the Gossamer Condor, designed by MacCready, which won the Kramer prize that defines the specifications of such flights. Prior to the Gossamer Condor, many other attempts failed because they were trapped by the limitation of conventional airplane configuration which could not yield a satisfying power weight ratio, despite the availability of newly discovered, exotic materials, such as graphite fiber, the mylar sheet, etc.

MacCready approached this problem by recognizing that a strong athlete can only deliver 1/3 horse power on a continuous basis. This, in his mind, was the most critical parameter and should be used as the starting point for innovation instead of conventional airplane configurations. Among his "building block" parameters, he included the fact that for a flight machine, power is proportional to the cube of the speed, and the wing area is inversely proportional to the square of the speed. This means that for a 1/3-horse power flying machine, he should try a very low speed, such as below 10 miles per hour, with an enormous wing area.

Also in his collection of building block parameters was one associated with the hang glider which carries the weight of man and machine at a very low wind speed. This is accomplished by the "deep truss" structure of the plane, which involves central, vertical tubing to suspend the loading of the wing with a series of thin piano wires. This structure is not "clean" from the point of view of its drag coefficient, but, according to MacCready's building block parameter, this is not important, because drag force is also proportional to the square of the speed and is not a problem when flying below 10 miles per hour. Adopting this simple concept and following it up with hundreds of minor, iteration steps, MacCready became the first to design a manpowered, flying machine. The Gossamer Condor is now exhibited at the Smithsonian along with the Spirit of St. Louis.

7. THE COMMONALITY OF BUILDING BLOCK PARAMETERS

The purpose of identifying a parameter is to recognize its common characteristics which can be applied to a class of problems. All technological parameters have this general criterion. However, in engineering science the greatest range of commonality is established by mathematical modeling, usually in the form of differential equations, which has the beauty of being indifferent to the physical parameters it represents; for instance, in wave theory, the same form of differential equation is applied to the electro-magnetic field, gas, or fluid media.

Fluid mechanics, for instance, is oriented toward application but is still very flexible. It is applicable to the study of current patterns in the oceans or the blood flow through the artery. It spans situations as large as the formation of the galactic spiral arms and as small as the movement of a sperm swimming toward the egg. Clearly, the configuration of an airplane is an incidental innovation, if viewed from the vantage point of fluid mechanics. This is one reason that it is very difficult to innovate from the parameters of engineering science. The concept of building block parameters represents one added step in developing technology, through system configuration. It must retain the commonality properties so that it is not unique to only one particular configuration, yet it should be novel enough so that people will not consider it standard practice.

One of the outstanding innovations in the heyday of aeronautics was the area rule introduced by Whitcomb of NASA. The area rule is an innovation which states that the sudden protrusion of the wing from an airplane fuselage should be compensated by a reduction in the cross-sectional area of the fuselage in order to reduce the aerodynamic drag. Traditionally, an airplane is designed with the wing accommodating a streamlined fuselage. Both wing and fuselage are studied separately and then joined together; here the law of partitionability (also in Chapter 3) applies very well, as does mathematical modeling of an aerofoil in a two-dimensional section, and seems to provide the primary guide for the aeronautics designer.

Whitcomb conceived of the area rule parameter for fighter planes by examining the potential flow pattern around the three-dimensional airplane body, visualizing in his mind's eye, the squeeze between the streamline pattern of the wing and the fuselage. His area rule primarily stated that for the middle section of a fuselage, the cross-sectional area should be essentially constant (Figure 2.4). The sudden protrusion of the wing from the fuselage disturbs this constant cross-sectional area,

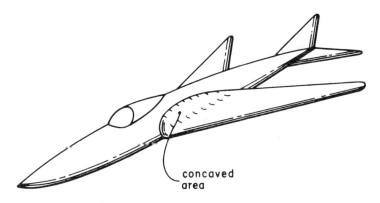

concaved
area

Figure 2.4. Aero rule concept for supersonic aircraft.

thereby necessitating a reduction in the cross-sectional area of the fuselage to compensate for the protrusion of the wing to minimize disturbance. The resulting shape of the fuselage is a "reverse bulge", similar to the familiar shape of a coke bottle. Experiments have confirmed his intuition that a significant gain in the maximum speed was achievable.

The ability to visualize the potential flow pattern, even in a crude way, around a three-dimensional, streamlined body gives an innovator additional freedom to innovate, because without this knowledge one tends to be bound by the two-dimensional concept. Whitcomb had that ability and identified his parameter as the "area rule." By assigning the name area rule to his concept, Whitcomb made it a conceptual parameter which carried a certain commonality, applicable to other similar situations.

One such similar situation exists in the design of the bow of a ship. A ship has a single, streamlined body subject to two forms of flow pattern. The submerged portion of the hull is a simple, streamlined body similar to that of a fuselage in subsonic, potential flow; accordingly, it requires a blunt nose corresponding to a subsonic airplane fuselage. On

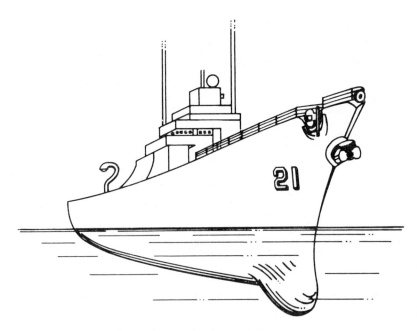

Figure 2.5. Utilization of underwater bulge to reduce drag.

the surface of the sea, the wave cutting phenomenon of the bow is similar to that of a supersonic body which dictates the need for a V shaped tip. The pressure distribution is characterized by the height of the wave propagating sidewise and downward to squeeze the streamline pattern of the submerged portion of the bow; this wave-generating behavior of the bow constitutes the greatest portion of energy loss to the ship.

One outstanding innovation in ship design was extending the submerged, blunt nose of the bow forward a considerable distance, followed by a "reverse bulge" under water but near the wave cutting edge of the bow at the surface. In so doing, the pressure gradient created by the submerged portion of the bow compensated for the pressure pattern created by widening the bow at the surface behind the sharp edge of the bow (Figure 2.5). In this new design, the added length of the submerged portion of the bow added more drag; however, the reduction in energy loss from minimizing wave generation at the surface is advantageous enough to justify the drag. The compromise between the two portions of the bow is indeed a generalization of the area rule, where widening of bow behind its knife edge is compensated by the "reverse bulge" of the submerged portion.

8. ACQUISITION OF BUILDING BLOCK PARAMETERS BY IDENTIFYING KEY PARAMETERS OF NOVEL DEVICES IN OUR DAILY ENCOUNTERS

For practical purposes, all current innovations are modifications of past innovations. The old expression, "reinventing the wheel" really happens all the time: We invent by incorporating refining improvements into existing devices, and these refining improvements are not usually entirely new concepts; therefore, identification of the parameters of a novel concept in an existing scheme, so that a similar concept may later be adapted to other schemes, is the first stage of innovation.

In the spirit of "reinventing the wheel," one recent "modification" of the wheel itself is the steel-belted, "radial" tire introduced by Michelin. This "modification" has a potential multi-billion dollar economic impact on modern society, because it doubles the life span of the tires; it also significantly improves gasoline mileage and traction in snow and on wet surfaces compared with the "bias reinforced" tire. While the average person might buy the tire simply for economic reasons, an innovator would seek the dominant parameters which make the radial tire work so well and would store these in his "bag of tricks" to be used for another application.

One important configuration in the construction of all tires relates to the layout of the reinforcing filaments. The tire and its reinforcements have the following dominating parameters: strength and rigidity in various directions; geometric shape under pressure; ability to transmit shear and lateral loading; and, finally, a "footprint" when making contact with the ground. The conventional tire was designed primarily with strength in mind. The reinforcing fiber was placed principally from one rim to the other (Figure 2.6), so that the geometric shape of the inflated tire resembled a doughnut. This structure was most efficient from the point of view of stress. The belted tire is reinforced at the extreme outside diameter (Figure 2.7) and along the tread of the tire, with emphasis on the longitudinal rigidity of the reinforcement. The geometric shape of the tire, when inflated, resembles a section of a cylinder (Figure 2.7).

The footprint of the doughnut shaped tire is elliptical, and a good deal of lateral rubbing between the tire surface and the ground is involved within the footprint when the doughnut shaped tire is rolling. On the other hand, with the "cylindrical" tire, the footprint will be rectangular, and ideally there will be no lateral rubbing of the rolling tire against the road surface. Lateral rubbing is the cause of all deficiencies in the

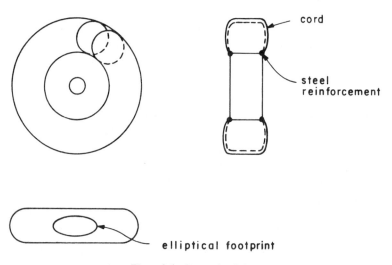

Figure 2.6. Conventional tire.

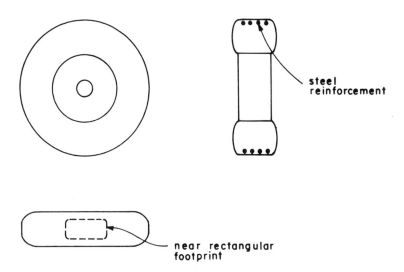

steel
reinforcement

near rectangular
footprint

Figure 2.7. Radial tire.

"bias" tire. This fact, and the remedy provided by the radial tire, represent valuable information to be stored as building block parameters in the innovator's "bag of tricks."

9. A POSSIBLE TRAINING PROGRAM USING PARAMETER ANALYSIS AS A METHODOLOGY FOR INNOVATION

Professional engineers can be trained to make design optimization if drilled in a certain, fixed design procedure. This is because conceptually, for a specific configuration, there usually exists a unique solution and a well-defined path for seeking it. Researchers in engineering science learn their methodology by serving as apprentices and working in a well-defined area. In an innovative company, the leader provides guidance to his juniors, and here too the master-apprentice relationship works well. Unfortunately, however, not too many industrial firms are endowed with capable innovation teams. After making an analysis of Figures 2.2 and 2.3, one may even wonder whether overtraining in professional engineering and research in engineering science inhibit innovative thinking. On the other hand, by simply exposing the college student to product development, there is a tendency toward superficial contact with trivial experiences.

Figure 2.3 illustrates the possibility of developing a wide span of new building block parameters which, if properly assembled, could be used as a structure for training in innovation. Skill in using parameter analysis to generate new ideas should be applied along all five stations of the cornucopia diagram in Figure 2.1, thereby adding dimension to the use of parameter analysis methodology.

The essence of this statement is that even though the goal of every innovation is a final product with a target specification, selling the conceived idea starts immediately after it is germinated, because, in order to carry the development program from station III through IV to V (Figure 2.1), a continuous expansion of funding support is needed. At each step, some selling is required to prove that the downstream risk is well thought out, with the remedy well planned to justify additional funding. All of this could be carried out effectively by parameter analysis.

Parameter analysis is useful not only for generating raw, innovative ideas but should also be applied to all steps necessary to bring the idea to fruition effectively and efficiently. After all, the final cost of the product depends upon the development cost of its realization. The leverage at each funding stage depends upon how much expenditure was required prior to that stage to reduce the perceivable downstream risk factor. The parameter analysis approach may also be viewed as identifying key parameter which will shorten the path to design computation and the dominating parameters which may warrant experimental verification to avoid trouble.

Even in the simple cat-chasing-cat game described above, it was shown that the solution could be quickly established after the key parameter was identified. Considering another example of the effectiveness of parameter analysis as an aid in design computation, a few years ago, the engineering team in a large aircraft company faced a space vehicle problem involving the undesirable, oscillatory motion of the vehicle when in orbit around the moon. The vehicle carried a deployable antenna boom about 40 feet long and went into an 8-minute oscillation period when it entered the sunshine side, quieting down in the shade. After 250,000 dollars worth of computer time dealing with some very complicated equations, the structural engineer in charge of the analysis was still puzzled. A parameter analysis approach was then introduced by dividing the boom into three sections, with the outermost section treated as a "lumped" mass, the inner most section as "lumped," elastic energy storage, and the midsection representing the dynamic energy coupler with the radiation energy of the sun. After the dominating parameters were thus identified, the 8-minute period was quickly verified and a cure suggested by minimizing the coupling between the system and the energy source. The fundamental rule about an oscillatory system focuses on the existence of potential energy storage paired with kinetic energy storage having a 180-degree phase change. In a cantilever system, the root of the beam plays the dominant role in potential energy storage, with the tip playing the dominant role in kinetic energy storage.

In general, an oscillatory system can be excited only when a force vector is applied in phase with the velocity vector in order to inject energy into the system. In this respect, the middle section is representative of an effective energy input coupling. This type of lumped parameter treatment often provides a quick check of dominant behavior. While it may appear to lack rigor to serious-minded engineering scientists, in making a parameter analysis during the innovative phase, one cannot afford to be too rigorous.

The difficulty faced by the structural engineer in the above example was largely due to the fact that he was trained narrowly as a researcher in one particular discipline and was not familiar with the dominant issues in other disciplines. Conceivably, if broadbased knowledge in building block parameters is assembled and taught in the university, it would serve as an excellent medium to unite various, technological disciplines and prepare future, industrial innovators and leaders.

10. PARAMETER ANALYSIS AND PRODUCT DEVELOPMENT

In the above section, it is assumed that a wide collection of dominant building block parameters of various engineering disciplines would make a sequence of excellent lecture subjects to prepare students to be innovators and industrial leaders. However, real experience can be gained only by participating in product development through the successive application of parameter analysis as outlined above. A typical checklist for a self-evaluation program of innovation development includes:

1. Specify the objective of the invention.
2. Identify the physical laws governing the achievement of the objective.
3. On the basis of these physical laws, identify the physical parameters that limit the performance of the invention.
4. Formulate the simplest, analytical model incorporating these pa-

rameters, and use this model to determine the appropriate range of these parameters and the sensitivity of the invention.

5. If the analytical model is too difficult, then try a graphical model, a three-dimensional model, or a numerical-analytical-graphical model.
6. Identify the "non-physical" constraints (by a similar parameter analysis), such as cost, methods of manufacture, materials, health and safety standards.
7. Using the results obtained from the model, design the simplest device (or process) that meets the objective within the imposed constraints.
8. Use the analytical model to evaluate performance of the design, and optimize this design within the physical and economic constraints imposed by the situation at hand.
9. Build a prototype to evaluate performance of the invention and to insure that the appropriate set of parameters was chosen initially.
10. Iterate the design until the simplest design, consistent with the imposed constraints, has been obtained.

11. VARIOUS ASPECTS OF INCENTIVE AND MOTIVATION

Even though innovation is a basic, human instinct, one's inner drive must be triggered by an external incentive. Many scientists are motivated by curiosity about the ever more fascinating natural frontiers, but to some degree social recognition, such as the Nobel Prize, provides the additional push. On the other hand, material reward is largely responsible for motivating innovators. The U.S. has been known as the haven of innovation in the past, because reward has been generous due to abundance of natural resources and the social/economic system which encourages private enterprise for personal gain. These incentives are diminishing while the need for innovation increases. Among several possible reasons, the most important is that resources and means of waste disposal are shrinking, while the world population is growing.

Rationization of the innovation process in order to structure it for teaching in universities is thought to be one possible way of minimizing the failure involved in developing ill-prepared concepts, the disappointment of investors who then hesitate to put their capital into circulation, and the quantity of talent which is ruined for lack of proper guidance. But teaching innovation without providing adequate incentives and motivation in appropriate forms for those involved in the development of the process is like creating a body without a soul. The individuals and groups to be considered in terms of incentive and motivation include:

- The student
- The faculty
- The university
- The industry
- The government

To satisfy the needs of all these parties, a university-based innovation center which encompasses a group of faculty specialized in developing parameter analysis methodology for technological innovation is considered. The faculty and staff will perform research work in developing a broad range of building block parameters which can be used as material for classroom teaching in several, possible subjects. Also affiliated with the center is an innovation co-op, structured as a charitable, non-profit, organization at arm's length from the university system. The co-op is to be staffed with enough professionals to perform product development work in a wide variety of technologies. However, all product development processes are to be reviewed using parameter analysis methodology to provide material for teaching and to allow the students to participate at various, appropriate stages.

In developing new products, the product itself is secondary to polishing methodology for improving the cost effectiveness of the development process. Because of this higher level, intellectual goal, the result should be worthy of publication and give faculty and staff a vehicle for gaining academic recognition—this is the kernel of the incentive needed for building up a new breed of faculty for teaching innovation.

While students are encouraged to be innovative, it is far more important that they be guided through the various stages of the product development procedure and sensitized to parameter analysis methodology than that they become unique in conceiving and developing their ideas. Thus the successful development of a student's own idea is incidental and not the goal of training. Furthermore, in the university environment, the student can be motivated most effectively by short term achievements in learning parameter analysis methodology when the teaching materials and the teaching format are well developed. The student's incentive will derive inspiration from working alongside the masters on real world, innovation problems.

The objective of the center must be to strive for excellence in the ability to develop a product of higher caliber than that of the average, industrial firm. If this goal is accomplished, the center should attract industry's interest and bring a wide range of industrial problems to the university for solution. Since the primary goal research in the use of parameter analysis methodology to improve the cost effectiveness of product development and to educate, the benefit to industry as a whole should be much broader than the few, new products which represent byproducts of the teaching process. For this reason, this kind of operation should not be considered by industry as unfair competition.

From the university point of view, establishment of an innovation center with a curriculum based on knowledge of various aspects of parameter analysis and building block parameters, which cuts across all technological and management disciplines, is indeed a unique and daring undertaking. Such a center would not only fill a void and meet a social need but represent a balanced teaching program, offering innovative students an opportunity equal to that of students concerned with careers in research, around whom the curriculum has hitherto been designed.

The conceptual diagram illustrating the information flow pattern from natural phenomena (at the top) toward product specification (at the bottom) also identifies three zones of laboratory activities. These three zones are marked on the diagram in Figure 2.3. The first zone represents the natural science research laboratory, which deals primarily with natural phenomena and touches the borders of technological parameters; in the middle is the engineering science research laboratory, which deals primarily with engineering science and runs from the edge of natural phenomena to product configuration; the third is the proposed innovation co-op type laboratory, which is concerned with configuration innovation and touches the edge of technological parameters and product specification. Touching the edge of the zone of specification without enveloping it completely, indicates that this type of laboratory activity is concerned with the scope of specifications but makes no attempt to fulfill them through product production and marketing, represented by station v of Figure 2.1. A fourth zone may be added, however, to represent the function of industry which would encompass the zone of both product configuration and specification. Certain industries with strong research centers might try to cover the complete range from natural phenomena to product specification, including all kinds of laboratory activities.

Its close resemblance to an industrial operation makes the innovation co-op type laboratory easier to manage if it is set up at "arm's length" from the rest of the university system. The uniqueness and success of the co-op operation hinges heavily on the educational function it serves, and the excellence it can achieve through that function. Due to the co-op's position at arm's length from the university, and the high level of technological innovation it can accomplish through collaboration with the teaching function inside the university, there is an excellent chance that the co-op will be financially viable and educationally unique and functional; both these factors should be an incentive for the university to give the concept serious consideration.

At the end of Chapter 1, the role of government regarding innovation was analyzed, and it was determined that individual product innovation is motivated by personal gain and is therefore the function of the private sector; however, improvement of the overall innovation process is of supreme, national interest and should be supported by government. The initial five year N.S.F. program for the study of the innovation center concept was a step in the right direction. The N.B.S.–D.O.E. energy related invention evaluation and development programs are all excellent policies. However, some strengthening and improvement of this type of program is definitely needed.

3
INNOVATION PROCESS
AS APPLIED TO
C.A.T.* SCANNER

1. BACKGROUND

Tomography is an x-ray projection system which is used to get a cross-sectional (or axial) view of the human body from the information gathered around the periphery of a particular section of the body in response to the x-ray beaming across the body at that section (Figure 3.1). With this machine, doctors can pinpoint the area of malignancy with an accuracy hitherto undreamed of. This is probably one of the best examples of the use of modern technology to provide direct, social benefit. The recent surge in the need for advanced medical equipment has induced several industrial giants to develop the C.A.T. scanner. Each is trying to overcome the problems associated with ingenious schemes, but the price tags are still quite high (in the neighborhood of ½ to 1 million dollars per system). President Carter's administration, believing that rapidly escalating medical costs are induced, in part, by the extensive use of exotic medical equipment, has put constraints on the purchase of the C.A.T. scanner by small and medium sized hospitals; even so, sales of C.A.T. scanners are still quite active, and the need for innovation remains.

The nature of a C.A.T. scanner was briefly described to Y.T. Li in 1975 by Arthur Chen, who led a team in developing one such machine at the General Electric Company. Then, in the summer of 1976, Y.T. Li

---

* Computerized Axial Tomography.

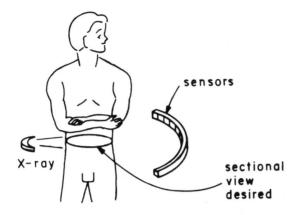

Figure 3.1. Conceptual representation of the C.A.T. scanner.

and his wife Nancy were entertained by Bernard Gordon (president of Analogic Corporation) and his wife at their seaside home in Magnolia, Massachusetts. The highlight of that evening was Gordon's account of how he had been awarded a contract from a multinational company to develop the data acquisition and computer portion of the C.A.T. scanner solely on the basis of his personal testimony of feasibility. The dramatic preparation made by Gordon for this contract involved a special trip to Purdue University to study mathematics with Professor Kak. Then, in discussions with Professor Cormack* at Tufts University, he was introduced to Johann Radon's 1917 paper "On the Determination of Functions by their Integral Values Along Certain Manifolds." After studying the paper and several idea-generating meetings with his senior engineering staff, Gordon was ready for his presentation, focused on achieving the fastest image formation offered by the industry. The closing of the deal hinged on Gordon's suggestion that if the client's own engineering team, after hearing Gordon's technical presentation, could not state clearly why the proposed system could not work, then Gordon would get the contract. Indeed, the client's engineering team did have doubts about Gordon's scheme but could not substantiate them. Gordon was then asked by the president of the multinational company to explain how he could be so sure that his group's scheme would work. Gordon responded that he had confidence in the technical aspects of the scanner and was ready to risk his reputation for it; with that, he got the contract.

At an advisory board meeting of the M.I.T. innovation center in the

* See footnote on p. 53.

fall of 1976, Gordon told Y.T. Li that his group had just shipped the prototype of the new C.A.T. equipment. The most remarkable part of this development work was that Gordon not only accepted a very difficult challenge, but actually honored his commitment to complete the work in record time (about six months). In terms of reward, Analogic stock quadrupled within one year, making it one of the fastest growing industrial firms in 1977.

In developing the methodology for teaching innovation, the feat accomplished by Analogic appeared to be a most interesting example. It was fascinating and dramatic to hear Gordon describing his bold approach in negotiating the contract. The fact that he got the contract without having his client achieve a thorough understanding of his proposed scheme makes this technological innovation even more intriguing. Yet it is the author's belief that most elegant innovations can be understood quite easily by trained individuals after the basic logic is identified. Even more inspiring would be an explanation of the choice of the path among alternate options to reach the final solution.

The following text is the result of several sessions with Bernard Gordon and his colleague Dr. John Hoff, when the author was led through a series of thought labyrinths. The key parameters of this innovation were finally identified and could be presented neatly as a technical paper, however, that would defeat the purpose of the present task. Instead, a step by step account of the thinking process (largely the learning process of the author) is herein presented, with the hope that the reader will emerge as elated as the author by appreciating that there just may be a methodology for learning innovation.

2. FIRST INTERVIEW WITH BERNARD GORDON

The first interview with Bernard Gordon was conducted on December 2, 1977 at the Analogic Corporation in Wakefield. The general principle and the common problem of the C.A.T. scanner was quickly identified. In concept, the information received by each sensor of Figure 3.1 at a given instant is the remainder of the x-ray beam of a given strength that penetrated the body. The accumulated opaqueness along the path of the light ray is supposed to be a known function of the ratio of transmission; from an array of these functions, the cross-sectional diagram is to be generated. As an example, in a simple 2 x 2 matrix of a cross-sectional area, as shown in Figure 3.2, the peripheral data available are S_1, S_2, S_3

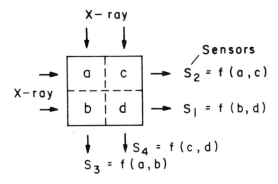

Figure 3.2. Determination of the cross-sectional view from laterally projected information.

and S_4. The relationship of four patches of the sectional view to the information a, b, c, and d is supposed to be known, as represented by the four equations:

$$S_1 = f(b,d)$$
$$S_2 = f(a,c)$$
$$S_3 = f(a,b)$$
$$S_4 = f(c,d)$$

With these four equations, it should be possible to solve the four unknowns: a, b, c, d. While a simple mathematical concept is easy to comprehend, solving simultaneous equations is, unfortunately, not a simple chore; furthermore, when the number of unknowns becomes larger than 100, the difficulty increases rapidly thereafter. For example, if the picture is to be constructed by a 500 x 500 matrix grid, there should be 250,000 simultaneous equations to solve. It is very likely that such a computation would strain the capacity of any practical computer.*

This first meeting was a casual one, and conversation dealt with the philosophy of training young engineers. Regarding the C.A.T. scanner, Gordon covered a wide range of topics in a rather random fashion; the following issues represent a summary compiled by Li:

1. The first electronic Tomography was introduced by Hounsfield† in England; however, the basic tomographic concept was conceived

---

* It is believed that some C.A.T. scanner systems do use the simultaneous equations approach. The problems mentioned here might be overcome through innovative computation programming.
† Cormack and Hounsfield shared the 1979 Nobel prize in psychology and medicine.

of as early as 1902, not long after the x-ray machine was introduced.

2. Johann Radon's 1917 paper "On the Determination of Functions by Their Integral Values Along Certain Manifolds" provided the key to the underlying mathematics of his system; however, Gordon commented that it was also possible to understand the C.A.T. scanner by examining it physically, instead of mathematically, though initially it did not seem to be too easy to do.

3. By manipulating pairs of transparent overlays, Gordon created a variety of Moire patterns. This type of approach is quite appealing to the methodology Li and his colleagues are trying to develop —Visual aids arranged to represent certain key parameters can usually help the innovator understand the dominant relationship between them. At that time, Li had not established precisely how to use the Moire pattern to understand the C.A.T. system.

4. Gordon introduced the transmission factor λ of material penetrated by x-ray as:

$$\lambda = \log \left(\frac{S}{R}\right)$$

where S is the signal received by the sensor, and R is the radiation strength.

Along with this equation, he emphasized that the real information is in the number of x-ray photons being blocked by the body and not in the number of photons which penetrate the body.

5. Emphasis was then placed by Gordon on the need for sensors with a million-to-one dynamic range; "Information Theory" was cited in support of this need.

6. On the blackboard of his office, gordon sketched a checkerboard to illustrate a typical test pattern. Around the edge of the checkerboard, he added a random, wavy line to represent the output signal from the array of receivers when the revolving pattern is illuminated by x-ray. Through the use of some aspects of the "Fourier transformation," including the "Gibbs" effect, an image of the checkerboard pattern was resolved from the wavy signal.

7. A lengthy, mathematical rendition by Dr. Dobbs, a physicist employed by Analogic, was credited with solving the spatial problem in the transformation. Leo Neoman had developed the com-

puter, Hans Weedor the data acquisition, and others had performed advanced mathematical and programming work.

A patent application, almost as thick as a telephone directory, bearing the names of Analogic staff, was also shown to Li, though he was not allowed to examine these documents because of Analogic's contractual restrictions.

8. Knowing that the Analogic system was based on a fan-shaped light beam instead of a parallel beam, Y.T. Li wondered whether that were in itself unique; the answer was no.

As the departure time drew nearer, Y.T. Li expressed the thought that even though a considerable amount of information had been gained from the interview, he had still not been able to identify the key issues underlying Bernard Gordon's innovation and suspected that Gordon had probably not developed his concept from the lengthy, mathematical derivation but from some insight into the physical nature of the system. Gordon agreed with this observation wholeheartedly but added that it was not easy to explain and suggested that a second interview might be helpful, promising that he would do some homework to retrace his thoughts before the next meeting.

3. SECOND INTERVIEW WITH BERNARD GORDON

The second interview was conducted Saturday morning, December 31, 1977 at Gordon's residence. As he had promised, a chart had been prepared, shown here as Chart 1 and Chart 2.

Chart 1, Item A: commitment to problem marked the decision of the management of an innovative company. To illustrate their total commitment, an artist's conception of the entire system was done before the design of the electronic circuit was shown to Li. Gordon was very pleased about the fact that the finished machine was not much different from the initial concept. Li was impressed by the depth of the innovation management illustrated on the chart but reaffirmed his main objective: to identify the key technical issues in Gordon's invention of the C.A.T. system. As Gordon's chart did not illustrate the technical concept, Li sought to outline his understanding of the issue (as a result of his own interpretation of the concept from their previous interview) in order to solicit Gordon's comments on specific, technical issues.

Chart 1. Schedule of Major Events in the Development of the C.A.T. Scanner

D	J	F	M	A	M	J	J	A	A	(month)
−2	−1	1	2	3	4	5	6	7	8	(sequence)
A/B	C/D	E/H	I	J/L	M/N	O	P/Q	R/T	U	(event)

A. *Commitment* to Problem and Schedule of Events
B. *Understanding* Problem
        Time
        Economics
        Competitive Efforts
        Tech. Goals
        Physical Basics
C. *Study* of Relevant Material
        Literature—Patent
        Technology
        Physics
        Mathematics
D. *Conceptual Systems*
    Conceptual Elements
E. Insight and Intuitive Direction
F. Concept (Subsystems)
G. Related Concepts
H. Integrated Concept
I. Sub-function Conceptual Pre-design
J. Sub-function Pre-design Assessment
K. Sub-function Comparative Analysis
L. Inter-engineering Promulgation of
    Understanding and Knowledge
M. Semi-Final System Design
        Test for Function
        Power
        Cost
        Space
N. Decision and Final Hardware Schedule
O. Detail Design of all Elements
P. Program, Contract, Pre-Test, Document
Q. System Test
R. Install
S. Refine, Rework, Improve
T. Final Documentation and Reporting
U. Production Test Equipment, etc.

Chart 2. Technology and Concept for the C.A.T. Scanner

- Convolution
- Pre-design Architecture—Physical System
- Mathematics
- Fundamental Information Signal Sources
- Limitations and Economic Constraints
- Realization of Fundamental Inter-limiting Relationship
- Radon's (Theory)
- Assessment of Scanning Geometry
- Mathematics of Singularities
- Display Technology
- Measurement Technology
- Detection and Detector Scanning (Measurements)
- Pipe Lined Computer Technology
- Conversion Technology
- Coordinate Transfer Matrix Concept
- Pre-interpolation Concept
- Logarithmic Analog/Digital Concept
- Wide Dynamic Range Concept
- Low Dose Concept
- Real Time Image
- Low Cost Computing System
- Low Noise—Low Dose

Y.T. Li stated that after he had given some thought to the problem, it occurred to him that a sectional x-ray system used to obtain a picture along the body axis is done mechanically by projecting x-ray beams rotating about a "falcum" plane as shown in Figure 3.3, so that all out-of-plane images are blurred. Gordon remarked, "That is what we were doing," and proceeded to draw a "sharp peaked" curve (shown in Figure 3.4) to illustrate the resolution to this kind of approach. Li then commented that he also recognized that the resolution of the scheme was a 'reciprocal" function but that he was not sure whether this function would provide necessary contrast.

Gordon answered that they had been bothered by the "infinity" at the origin of the reciprocal function. Fortunately Dr. Hoff, a former mathematics professor from Tufts University, had applied for a job at Analogic and, at the time of his interview, revealed that his specialty was solving functions involving infinities. Gordon briefly sketched a transformed reciprocal function (shown in Figure 3.5) to illustrate how Analogic had dealt with infinity but declined to go any further or to let Li scan Dr. Hoff's paper because of confidentiality factors. As for synthesis of the signals to form the desirable pattern, Gordon referred back to the checkerboard he had used as an earlier example and commented that signals had been taken from sensors and used to make a spatial Fourier Transformation which, unfortunately, was too complicated to comprehend outside of a mathematical context.

At the conclusion of the second interview, both Gordon and Li agreed that, aside from talent or skill in innovation, the satisfaction of one's ego had a great deal to do with striving for success. Gordon's hard-driving spirit was well illustrated by the tasks outlined in Charts 1 and 2 and further demonstrated by the fact that he had spent the previous Thanksgiving Day scanning through four thousand patents to prepare for a litigation case. Gordon admitted that this kind of driving spirit was influenced by his own egotism.

Li was well gratified by confirmation of his belief that an instant C.A.T. system must be one capable of *summing up all information passing through each data point;* this was the first key parameter that he had been seeking to establish for this particular system from the standpoint of technological innovation. While many innovators are skillful enough to recognize the key technological issues, few can get the job done as spectacularly as had Gordon. Gordon's feat reminded Li of the comment made by Al Kelly, Dean of the School of Management at Boston College, who was known as the expert in venture business investment. In answering Li's question, "What does a venture business investor go by in evaluating an entrepreneur who has come to him with an innovative scheme?" he remarked that he would invest only in the person and never in his scheme. Thus, while this interview began with the objective of identifying the key issues of technological innovation, the responses elicited from the innovator leaned more heavily toward his attitudes as an entrepreneur.

4. INITIAL ATTEMPT TO INNOVATE ON THE CROSS-SECTIONAL PROJECTION SYSTEM

As Y.T. Li pulled his car out of the Analogic parking area after his first interview with Bernie Gordon on December 2, 1977, he was quite confused. He had hoped to get a simple description of how Gordon'

C.A.T. scanner system worked, but instead he had been showered with fancy, mathematical terminologies and an itemization of management constraints. On the other hand, he had been assured by Gordon that one must innovate with clear insight into the physical properties of the system, instead of being entangled with mathematical formulations and system details. As Y.T. Li's car merged with Route 128 traffic, the steady act of driving allowed him to examine the key parameters objectively, instead of guessing at Gordon's system configuration by piecing together his words.

The first parameter that came to mind was the x-ray beams which penetrate the body. He appreciated at that moment Gordon's comment that the needed information lies in what is being absorbed by the body tissue and not in what penetrates. However, he also realized that if the x-ray beams radiated with known, initial strength, then one could consider either the portion of the beam being absorbed or what penetrates as the information. A mental picture of the penetrating rays began to evolve, and one ray was observed as being absorbed by the various body tissues along its path. The location and opaqueness of these tissues were really the needed information, and, given the output signal of this one ray, only the sum of the opaqueness of the intersecting tissues was known.

Li's mind then drifted toward consideration of a second ray which intersected with the first, and, vaguely, it appeared to him that the information at the intersection point was common to the output signals of the two rays. Encouraged by this interesting little "hint," several more rays intersecting at the same point were considered. The fact that each of these rays contained the same ingredient of information about the tissue at the intersection point must mean something. One temptation was to try to add them together to see whether that would bring the common information out by mutual reinforcement of all the intersecting rays. Suddenly, he recalled the scheme used by the longitudinal sectional x-ray machine, which allows a bundle of parallel x-ray beams revolving about a certain sectional plane in the patient's body to be used as focum (Figure 3.3), so that the image of the body tissues on the focum plane will be reinforced on the film, while images of other planes will be blurred by blending with each other. Fortunately, Li had been fascinated enough by this scheme when he first heard about it years ago to make a mental analysis of its key parameters; it thus constituted a building block in his "bag of tricks." Now, while driving along Route 128, he quickly reconstructed in his mind the basic configuration of the longitudinal sec-

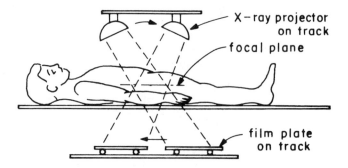

Figure 3.3. Image enhancement of a longitudinal x-ray projection.

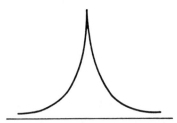

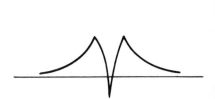

Figure 3.4. Image contrast as a result of first order image enhancement.

Figure 3.5. Shape of modifying signal (proposed by Gordon).

tional x-ray machine and realized that in that machine, even though the film were exposed to information on all sections of the body, as in an ordinary x-ray machine, only the information near the desired section was in focus to make the image of that section stand out against a blurred background.

Encouraged by the existence of the longitudinal sectional x-ray machine, which apparently produces an acceptable picture, Y.T. Li began to wonder whether a similar principle could be adapted to construct an axial view sectional machine. He soon realized that, in the longitudinal sectional machine, the body section selected for picture taking and the film are parallel to each other thus making direct projection with parallel beams a simple matter. The equivalent situation for the axial sectional machine would be a design which placed the film side by side with respect to the selected body section where, somehow, each ray of information that penetrated the body section would be laid upon the film at the corresponding position, as shown in Figure 3.6. (This logic can be explained more clearly by a simple law of statistics, described later.)

In Figure 3.6, three rays *a-a, b-b, c-c* are shown passing through a body section. The corresponding information, a'-a', b'-b', and c'-c', is laid on the surface of the film, which is placed beside the body section. In so doing, the information at the three intersection points of the three information lines is mutually reinforced. Thus, if a sheet of beams sweeps across the body section while allowing the body section to rotate over an angle of say 180 degrees about its longitudinal axis, then every point in the sectional plane will be illuminated by a revolving ray. If all rays that penetrate the body are then allowed to be laid down sideways onto the film at the corresponding position, then a cross-sectional picture, equivalent to the longitudinal sectional picture, would be obtained.

The synchronization of the film with the body is easy, especially if parallel beams are used, as shown in Figure 3.7, where the x-ray projection system is shown as stationary, while both the body and the film are revolving in synchronized fashion. Of course, if it is desirable for the body to be stationary, a similar effect can be achieved through simple, mechanical gearing. Laying the ray down sideways onto the film requires some innovation; while Y.T. Li's car was approaching his own driveway, he discovered a crude scheme for achieving this objective. That scheme involved a thin sheet of transparent material in which many, fine particles are suspended. These particles would scatter the x-rays like the dust particles in the air when illuminated by a beam of

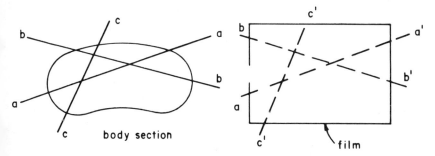

Figure 3.6. Image building by laying outcoming rays on film at corresponding positions.

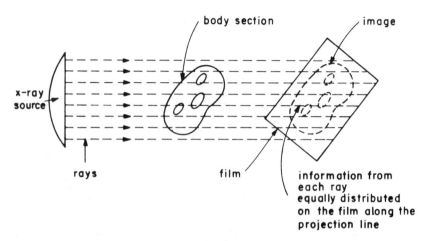

Figure 3.7. Parallel x-ray beam concept.

light in the night. Thus, when this sheet is placed over the film and receives the x-ray edgewise, the scattering of the x-ray will lay its information on the film, along the path of the penetrating ray. Li realized at that time that this is not an efficient system but was quite pleased with the innovation process which had occurred during the 20-mile drive from Analogic to his home in Lincoln.

5. REFINEMENTS OF THE CROSS-SECTIONAL PROJECTION SYSTEM

During the period before and after the second interview with Bernie Gordon, Y.T. Li made further refinements in the scheme outlined above. It is interesting to note that the initial phase of innovation conducted by many innovators occurred in just about the same manner as that described in the last section. "Intuition" represents a logical, thinking process which matches up some "building block" one has stored in his "bag of tricks;" an account of this process might help the reader to conduct his own exercise. The refinements that follow the initial innovation usually represent much diversified effort whose chronological sequence has no particular meaning, except from a management point of view, if the device is ready for development with expenditure versus business commitment at stake. For the present task, refinements were carried out primarily as a random, academic exercise. In retrospect, the various directions for refinement may be considered as system configuration and resolution study; these were then followed by a philosophical discussion, herein called parameter analysis methodology. The improvements in system configuration are summarized in this section.

Figure 3.8 shows a scheme for distributing the x-ray signal to the film on which the cross-sectional picture is to be projected. For convenience of illustration, the x-ray source and the rays are fixed with respect to the paper, while the "body" and the "film" are shown to revolve in synchronized fashion. The rays may be parallel or fan-shaped. The output signals from the sensors are connected to an array of signal conditioners, where appropriate output impedance and non-linear characteristics, such as logarithmic function, may be introduced. After "conditioning," the signals are distributed to a set of commutators, each of which is geometrically fixed in correspondence to the various x-rays they represent, thereby commutating with the brushes affixed under the film plate. Each brush corresponds to a spot in the picture of the cross-sectional view. Every time a brush moves across one commutator section, a "sample and add" operation results. The total accumu-

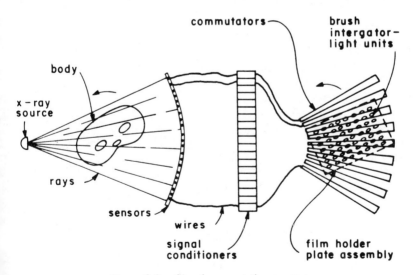

Figure 3.8.   Signal commutation concept.

lated signal of each brush after the body and film have completed half a revolution is then used to illuminate a picture spot, thereby exposing the film. It is interesting to note that in d.c. motor design, there are now "brushless," or "electronic commutating," systems. A similar, brushless scheme could certainly be used for this application.

Light Beam Commutating Method

Figure 3.9 is a system which evolved directly from the original concept shown in Figure 3.7, except that fiber optics are used instead of the light scattering particles, thereby improving efficiency and quality. The system shown in Figure 3.9 consists of a revolving body (for convenience of illustration) and a revolving film. After penetrating the body, the rays shine directly on a horizontal row of small windows, where an array of fiber-optic bundles terminate. Each bundle of fibers originating from each window is spread out to illuminate the film along one narrow wedge, which corresponds geometrically to the rays that penetrate the body tissue and shine into that particular window. In this system, the film performs the integration, as well as the nonlinear function, in exactly the same manner as in the conventional, longitudinal sectional x-ray machine.

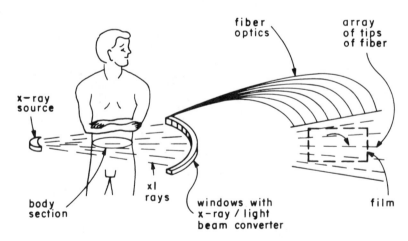

Figure 3.9. Signal commutation with fiber optics.

Computerized System

A computerized cumulating system may be conceived of by starting
with a simple system, such as one with:
- parallel beam
- constant speed

As shown in Figure 3.10, a certain spot A in the body section is illuminated by a group of rays with its shadow falling upon an array of sensors. The movement of the shadow is a simple sinusoidal function when
the two conditions stated above are held. A group of time-delays is used
to couple the sensors to the corresponding spot A' of the film matrix, so
that A' will "see" the shadow of A throughout the illuminating process.
Likewise, another set of time-delays will be used to couple spot B' with
the sensors and so on for all the spots of the film matrix and the
influential sensors.

Additional modification of the time-delay function for a fan-shaped
beam or even non-uniform, rotational speed is certainly not hard to imagine, if operating in the geometrical instead of the time domain.

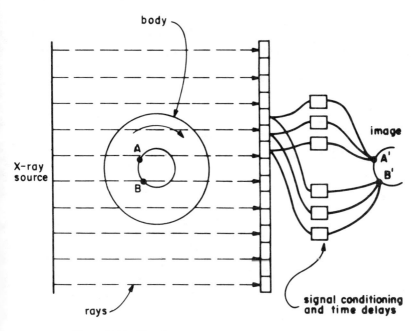

Figure 3.10. Signal commutation with electronic computer.

Figure 3.11 shows another version of the light beam commutating method, where the row of windows in Figure 3.9 are replaced by a cylindrical lense, so that the sheet of the x-ray that penetrates the body section is spread out into a diverging beam in order to expose the entire area of the film.

6. THE RESOLUTION STUDY

Developmment of all the various system configurations described in the last section was based upon one single principle of logic, that is, add all the signals of the rays that are crossing one spot in a body section, and use the sum as information for the corresponding spot on the picture to be constructed. Confidence in this concept was reinforced by the realization that it is the same concept utilized by the longitudinal sectional x-ray machine. However, prior to the second interview with Bernie Gordon, there was no assurance that this summation concept was adequate when compared with other methods involving powerful, computerized, mathematical formulation. Thus, to get a better feeling for the adequacy of the approach, a few simple tests were conducted. The first was the reconstruction of a single point.

Figure 3.12 shows the test sample with a single, solid dot situated at point A in specimen defined by the boundary line which is scanned by a

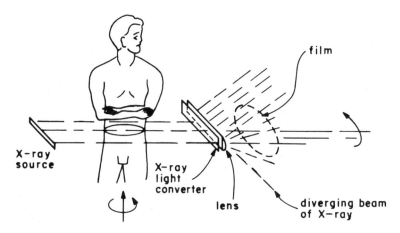

Figure 3.11. Commutation with a prism (proposed by Vander Velde).

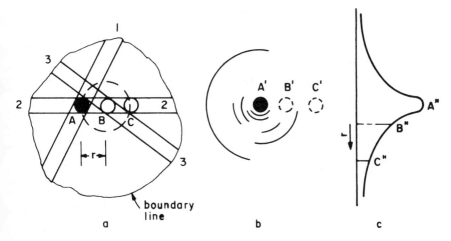

Figure 3.12. Concept underlying simple image enhancement.

revolving sheet of x-ray beams. Rays 1-1 and 2-2 represent those rays which revolve with spot A as center, with the sum of the signals corresponding to these rays being used to generate the picture image at spot A'. Rays 2-2 and 3-3 represent the rays which revolve with spot B as center, and likewise the sum of the signal corresponding to these rays is used to generate the picture image at spot B'. It is quite obvious that all rays that revolve with A as center see the solid spot at A. The total strength of the sum of these signals creates the dark core at A' in Figure 3.12b.

The rays that revolve with B as center only occasionally get the full signal strength when that ray also intercepts the solid spot A, as in the case of ray 2-2. The total signal strength for spot B', as compared with the signal strength at A', is equal to the width of spot A divided by the length $2\pi r$ of the dotted line circle. All spots around the same circle should get the same strength, so that the picture thus produced for solid spot A becomes a diffused picture, as illustrated in Figure 3.10-b. The strength of the signal is reduced in inverse proportion to the radius, except near the center, when the final size of the spot and the resolution of sensor are considered. Conceivably, the finer the sensor, the sharper the resolution.

The second test used the checkerboard pattern suggested by Bernard Gordon during the first interview. The test pattern selected by Li was a

4 x 4 checkerboard shown in Figure 3.13. The purpose of the test is to determine by graphic method the contrast of the picture image at points *a, b, c, d,* and *e* when each point is scanned by 18 rays 10 degrees apart, as shown in Figure 3.14. Figure 3.14 is drawn on a transparent overlay with its center placed at the respective points *a, b, c, d,* and *e* of Figure 3.13. A divider is used to measure the strength of each ray as it passes across the checkerboard pattern. For convenience, the cross area of the board is considered as "plus" and the shaded area as "minus." The strength of each ray is tabulated as shown in Chart 3. This test is very illuminating and encouraging. Before it was tried, a glance at the interlacing nature of the pattern suggested that there was no hope for its resolution, primarily because the eyes were attracted to the nature

Chart 3. Strength of the Rays Passing Through the
Checkerboard Pattern

shaded = −   light = +

	Deg.	a	b	c	d	e
0	0	0	0	0	0	0
1	10	− 7	−19	−18.5	−16.5	0
2	20	−15	−15.5	−12	− 8.6	+15
3	30	0	+ 7.2	+ 6	− 6.5	− 6.5
4	40	− 9	+ 1	−19	+33	−30.5
5	50	−14.3	− 6.5	+14.5	+32	−30.5
6	60	− 6.5	− 4.5	+ 5.5	+ 5.8	−11.5
7	70	+ 8	+ 3.8	−18	− 8	+15.5
8	80	0	0	− 4.5	−14	0
9	90	0	0	0	0	0
10	100	0	0	+ 8.8	+10	0
11	110	+ 8	+ 3	+15	+ 5	+17
12	120	− 6.5	− 4	− 5.2	−10	− 6
13	130	−21	−11.5	−33	−41.5	−40
14	140	−31	−21.5	−37	−44	−40
15	150	− 9.2	− 6.5	−10	−10.5	− 6
16	160	+ 2	+ 5.0	0	+ 4	+17
17	170	+14	0	+ 7	+ 9	0
18	180	0	0	0	0	0
		−87.5	−72	−62.4	−79.8	−106.5

of the principal axis where the signal strength of the rays was indeed zero.

The sum of the signal strength for the various points is given on the bottom line of Chart 3, with the peak e at the center of the square registered as -106.5, while the center of the edge b is -72, and the corner d is -79.8. In order to be symmetrical, the corresponding points on the cross area should have the same but positive value. The sectional view of the signal parallel to the principal axis is shown in Figure 3.15.

7. PARAMETER ANALYSIS APPLIED TO THE C.A.T. SCANNER SYSTEM

One of the organized methods for stimulating innovation was known as brainstorming, which involves a small group of people calling out random suggestions for evaluation. Y.T. Li's first interview with Gordon was, in some respects, like a brainstorming session: First, Gordon was bound by his obligation not to disclose any confidential material to an outsider, so that the information he conveyed was limited to peripheral, common knowledge; furthermore, as Gordon stated earlier, even his client's engineering team had difficulties in fully appreciating his orginial presentation. this could mean that every innovator has a code of mental building blocks for developing his own innovation, which is then doubly hard to convey without disclosing the configuration. On the other hand, random information provides the best beginning for innovation, because it activates free thought with an unrestrained mind. To sort out some useful information from a random pile and then to construct an innovation is somewhat like solving a jigsaw puzzle. Each piece of the puzzle must be *examined* for its special feature, such as the contour of the edge or the telltale of the printed pattern, which is a portion of the master design of the puzzle. This would then be followed by a trial fitting of pieces guided by one's experience.

The information to be sorted out for innovation is certainly much more complicated; it may associate elements in different types and forms, such as hardware, specific systems, conceptual models, mathematical formulations, economic conditions, producibility, user's taste, and many others. In the sorting process, each element is *analyzed* to identify its unique *parameters* which couple with other pertinent elements. This information sorting for innovation is called the parameter analysis process.

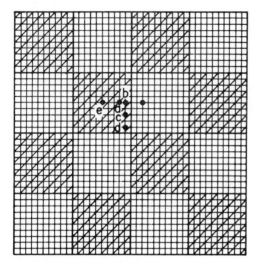

Figure 3.13. Checkerboard pattern used as a test of signal contrast.

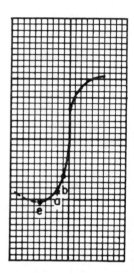

Figure. 3.15. Result of signal contrast of the checkerboard using simple image enhancement.

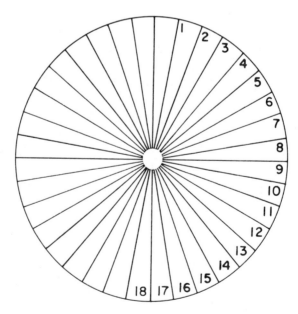

Figure 3.14. Orientation of rays used to test the checkerboard pattern.

In addition to the complicated nature of the various types and forms of elements to be fitted together, some of the information received by the innovator through the random process may be useless and detractive and must be recognized as such and discarded through parameter analysis. On the other hand, the crucial element is often missing from the new inputs and must be drawn from the innovator's own "bag of tricks" where potentially useful elements were "catalogued" according to the "parameter analysis" coding system he adopted.

A jigsaw puzzle is solved when the master design is completed; in innovation, there is no such master design or preconceived configuration to terminate the game. Most of the time, parameter analysis is carried out continuously in many iterative stages, such as initial sorting, trying out, reconstruction, bringing in new information, etc. Somewhere down the line, a novel configuration emerges with a promising, improved performance. At this point, many innovators jump with joy and start to "cut metal," so to speak. However, without a definitive master design as in the jigsaw puzzle, it is difficult to say whether a novel configuration is the best possible scheme. For instance, at first glance, the optical, cumulative scheme described in Figure 3.10 seems to be quite adequate as a low cost tomography system, and yet, the following question lingered on: "How does the Analogic system work?" "What is the needed resolution?" "Can a computer really improve the resolution?" and "What is the theory introduced by Radon?"

Answering these questions directly requires much further study and inquiry, which may be futile because of the confidentiality considerations at Analogic. For the present study which emphasizes using this exercise to illustrate the role of parameter analysis in education, it seems most sensible to conduct further parameter analyses on the philosophical level with the use of existing information.

The first important and interesting philosophical parameter of the system is the *partitionality* of the two main groups of physical parameters pertaining to the two information chains of the system. Essentially, the purpose of the C.A.T. system is to relate each spot on the body section to a corresponding spot on the film. To complete the coupling, two chains of information are involved: one deals with the geometrical coupling, while the other deals with the light transmission property of the tissue at each spot. All signals are initiated by the x-rays, but the information of the light transmission factor of each spot is carried *along the direction of the ray*, whereas the geometrical information of the spot is carried by the *lateral position of the ray*; these two parameters are orthogonal to each other and, hence, independent. From there on, the two chains of information remain independent, unless they are purposely mixed for beneficial reasons. For geometrical coupling, the commutating concept described in Figure 3.8 seems to be broad and general enough so that detailed innovation can be made at random without significant improvement in performance over the several schemes previously suggested.

To construct the image to represent the light transmission property of body tissue at each spot on the body section, the scheme conceived intuitively by Y.T. Li follows in fact a fundamental statistic theory. That scheme, described in section 4, involves summing up all the information from the rays that pass through a particular spot on the body section and using that summation to synthesize the picture spot for display. This idea occurred to Li, who assumed that each intersecting ray is influenced by the behavior of the same spot and hence their sum would mutually reinforce each other. Li's previous knowledge of the longituditional sectional x-ray machine provided solidification of his intuition, while the checkerboard test provided encouragement, and the confirmation by Gordon provided assurance. But all of these are various stages in the certification of an intuition: Only association with its statistical explanation finally gave supreme satisfaction, because with the probability or likelihood factor, a new vista was opening up which not only anchored the intuition to an absolute scale but also provided a direction for further improvement.

According to the theory of statics, each ray shining through a body is absorbed by the tissues along its path, which would give a uniform probability distribution function over the entire path relative to the point of absorption. This is so because it is assumed that there is no prior information on the pattern of the sectional view, so that the distribution must be totally random.

The transmission factor of each elemental block is defined by the ratio of output to input. By taking the log of (output/input), the transmission factor (or opaqueness) is linearized, so that its statistical nature can be processed according to established rule. Linearization means that for a multiple layer of light-absorbing material the sum of the log (S/R) of each layer is equal to the observed log (S/R) of the entire ensemble. If the distribution of the transmission factor of the various spots in a body is totally unknown (or random), then a second ray shining across the body section will also yield a uniform probability distribution function independent to that of the first.

If each of all the spots in the body section are subjected to the same number of revolving rays, then the maximum likelihood of the opaqueness factor (inverse sign of the transmission factor) of each spot is the sum of all the probability factors contributed by all the testing rays (assuming that the transmission factor is non-directional). Thus, according to the theory of statistics, the process described earlier was quite sound and cannot be improved for any totally unknown topology of the body section. However, in general, we are dealing with the human body composed of known, basic features, except for unknown, new growth. One might wonder whether there were some advantage to be gained from the known, common pattern. Indeed there is: the optimum estimation method developed independently by Kalman and Batton for space navigation during the 1960s can improve the estimation with a priori information. Professor Vander Velde of M.I.T. not only feels that this is an ideal application for optimum estimation, but because of the quantity research and computerization in this area, the adaptation of the concept to the C.A.T. scanner system should be quite rewarding, if it has not already been done; of course one must realize that the computational burden would be enormous.

Vander Velde also remarked that the equations for solving unknowns by matrix and determinants should yield the same solution obtained by the optimum estimation method. Mathematical correlation between the two does exist with the solving of simultaneous equations, frequently called the batch method, versus the recursive method, which characterizes the optimum estimation process. For operations where the information is collected continuously, as in the targeting phase of space navigation and the C.A.T. scanner system, the recursive method is certainly more appropriate than the batch method, because in the latter computation can start only after all information is collected, making the process slower and, accordingly, necessitating more information storage.

Based upon the above analysis, the following conclusions may be drawn.

● The optical commutating type, as shown in Figure 3.10, is probably the most economical scheme for a cross-sectional view x-ray machine. The picture quality should be about the same as the conventional, longitudinal sectional view x-ray machine (commonly known as Conventional Tomography according to Dr. Dobbs).

- The introduction of sensors and logarithmic functions to linearize the transmission factor and then to do the commutating and summing should yield a better resolution at additional cost.
- A computer may be used in conjunction with the aforementioned system to perform the optimum estimation. To be effective, a priori information must be provided. The greatest advantage of optimum estimation is the opportunity to provide a picture of comparable quality with significant reduction of x-ray exposure to the patient. A reduction of two to one of the exposure is quite possible, which is especially beneficial to the patient who must undergo repeated examinations.

8. ON THE IMAGE ENHANCEMENT METHODS

On March 15, 1978 Y.T. Li met again with Bernard Gordon and, for the first time, with Dr. Dobbs. Gordon complimented Li on having done some nice "thesis work" in deciphering the basic concept of the C.A.T. scanner and on his philosophy of innovation, described in last section. However, Gordon did not approve the schemes suggested by Li, including the use of the photographic film for integration, because it would not give the desired resolution. Between Gordon and Dobbs, and with the aid of the blackboard, the concept of the enhancement of the image of the Analogic C.A.T. system was described to L as follows:

Figure 3.16 shows schematically the conversion of a single information source (represented by the cylindrical post on the left) into an image with the horn-shaped density distribution function (shown on the right. The conversion is accomplished with the simple "summing" method which was described in Figure 3.12 and now is symbolically identified a an operator A. The intuition is, therefore, to insert another operator which, in combination with A, would reproduce an image with the same, cylindrical information density as the source. The function of operator B has the characteristics represented by the "folded cone" shape profile. But how to insert operator B by the convolution method in the information system was not readily apparent. As an analog Gordon suggested using an electronic equalizer (which is just a filter) improve the fidelity of an audio system. Gordon and Dobbs both f that mathematics is just a "tool" which is useful in carrying out a d tailed analysis of a conceptual system, such as the filter analogy abov

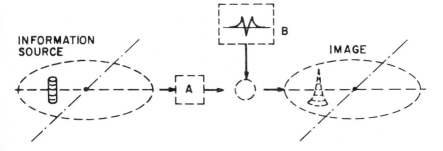

Figure 3.16. Image enhancing by convoluting the signal from system A with a weighting function generator at B.

Li concurred, but only after further deliberation and consultation with Professor Ho of Harvard University was he able to get a clear picture of the three-dimensional, convolution process Dobbs had tried to elucidate.

For most readers, the concept of using a filter to equalize the fidelity of a loudspeaker is quite straightforward, and it is indeed a good "building block" for innovation. The transformation of the frequency response concept to its equivalent, the spatial convolution, requires the following four concepts:

- Frequency response is characterized as a phenomenon in the "frequency domain," where the input signal is exhibited in its frequency spectrum, and the system behavior is expressed in frequency response. The output signal is obtained by operating the input signal with the system response by a rather simple, algebraic process.

- For linear systems, the frequency domain behavior and the time domain behavior are interchangeable. In time domain, the input and output are functions of time, and the system behavior is expressed as a *weighting function*. The output signal is equal to the convolution of the input with the system weighting function.

- Two-dimensional, geometrical convolution is the same as convolution in the time domain, with time replaced by geometric dimension. There is one slight twist in making the geometric based convolution as opposed to the time-based convolution, because in real world time, it can only advance forward, while in space no such constraint exists.

● Three-dimensional, spatial convolution follows the same rule as the two-dimensional, geometrical convolution; both convolutions can be illustrated easily as follows:

In making the two-dimensional convolution:

Use the weighting function to scan across the original signal from one end to the other, one step (corresponding to the pixel of the image) at a time, and use the center of the weighting function as the reference. At each step, multiply the two sets of adjacent values (the signal and the weighting function), and use the sum of all these products as the value of the modified function at the point of reference.

Figure 3.17 shows the original signal in the shape of an inverted V which is convoluted by two sample weighting functions to yield two modified signals. These sample operations are used to show that the V shaped, original signal can be modified into the approximation of column shaped profiles by two different weighing functions.

In making the spatial convolution, the weighting function is converted to the corresponding surface of revolution to convolve with the horn shaped, unmodified signal. In so doing, the reference axis of the weighting function should scan through the entire area, so that a new value for the modified signal is determined for every pixel point in the image plane following the same process described above. In principle, the sequence of scanning makes no difference; for instance, it is possible to attach a weighting function bundle (say 100 points) to each pixel (say 60,000 in total). Thus, every time the neighbor of a particular pixel which is being scanned receives an incremental incoming signal, it will be multiplied by the value of the corresponding weighting function and then added to that pixel. In this manner, the magnitude of the pixels of the modified image will emerge gradually while the x-ray is scanning, with the final image complete as soon as the scanning stops. Based upon the law of superposition for linear systems, Dobbs believed that what is effective to sharpen the image of a single spot should be equally effective for other patterns which may be represented by the aggregation of many spots. For instance, the linearity of the operation A of Figure 16 may be tested by observing that the image generated by the coexistence of two simple spots is equal to the sum of the images of two spots developed separately.

Image enhancement is a field by itself: The key parameters of an

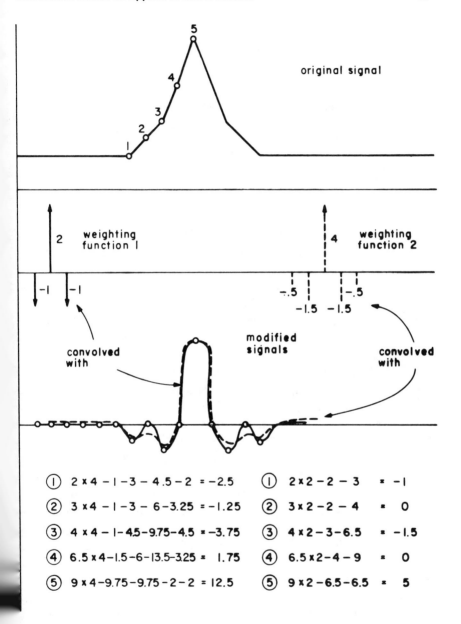

Figure 3.17. Simple example of the convolution of two signals (prepared by Ho).

the commonality between image enhancement and optimum estimation deserve further study, but, for the present exercise, brief mention without detailed analysis will suffice.

9. ENTREPRENEUR, FACULTY, AND STUDENTS

In this episode, we have seen the conceptual development of a sophisticated and socially relevant product which was to be marketed by a multinational company. Among the teams of engineers and experts related to this program, we observe an important role played by university professors; however, at the center of the stage is Bernard Gordon as entrepreneur. He is the one who assembled the team of experts to tackle the job; he is also the one who made the initial survey of existing knowledge in order to map out an approach for exploring the unfamiliar ground of the C.A.T. scanner. When he ran into difficulties, he consulted with, or hired, professors to verify or extend his initial concept. Then, when he was ready and his client willing, he set his team of experts to work following the detailed programs outlined in Charts 1 and 2.

At the university, students are trained to think like their professors. Some get industrial Co-op training and thereby develop skill as engineers. Although in basic science and engineering science, professors set the pace in research, in industrial operations, their input must be solicited; furthermore, engineers tend to be followers awaiting instructions. The entrepreneur gets satisfaction from seeing his creation—the product—accepted in the marketplace. He needs innovators to help him achieve his goal, but it is not his primary interest to train them to be innovators or entrepreneurs. The question remains, should the university attempt to, or is it even able to, train students to be future industrial leaders and innovators, who, in effect, become individuals who ask pertinent questions leading to resolutions which meet social needs, and who coordinate solutions to accomplish goals?

Maybe the unique characteristic of entrepreneurs and innovators is their ability to weave through the yarns of many specialties to form a fabric. They must interface with a wide spectrum of specialties and yet refrain from becoming generalists by indulging in quantities of diversified, but unrelated knowledge. Instead, they need to know what each branch of technology can do and the structure for obtaining data if necessary. Of paramount importance is an awareness of the process needed and means at their disposal for converting a scheme into reality

This chapter attempts to identify the interlocking parameters of various technological disciplines, including the relationship of the various technologies to the need which the final product is intended to serve. Skill in identifying the commonalities of the various key technological parameters is like manipulating yarn on the shuttle of a loom to weave fabric, which not only provides a tangible medium for carrying out the teaching function with regular students but is beneficial to professional engineers.

It would be even more effective for education if it were possible to have a product development organization, like Analogic, act as a teaching hospital affiliated with a university, such as M.I.T., so that the continuous interface between technology and management might be analyzed, observed, and practiced by students. Further development of these themes will take place in the remainder of this work.

4
SYNERGISTIC NATURE OF INDUSTRIAL MANAGEMENT AND TECHNOLOGICAL INNOVATION

1. THE PRIMER OF TECHNOLOGICAL INNOVATION

Over the past three decades, the ebb and flow of technological innovation have dazzled the industrial world. The spectacular moon landing, a result of the massive U.S. space program, led people in the sixties to think that the U.S. government was the primer of the innovative process in this country. Then there was the fabulous growth of industries, such as I.B.M., Bell Labs, and General Motors, which led some management experts to claim that "high technology" was becoming so complicated that only the giants could marshall the capital and manpower to make significant technological progress. However, other management experts, such as Professor Edward Roberts of M.I.T., maintain that small industry still has an edge on innovation; thus the pattern of small industry pursuing technological innovation deserves the attention of large companies, who can learn from and even emulate the smaller firm.

Each of these observations has its own validity, and this kind of study serves well its academic objective; This chapter is not, however, concerned with grouping theory or its equivalent but proposes rather, through an anatomical approach, to dissect the dominant factors which influence technological innovation in industry. Accordingly, let us take a close look at the role technological innovation plays as a given firm grows from small company to industrial giant and strives to maintain its leadership.

In a broader sense, the industrial operation invariably involves engineering and management. Conceivably, some engineering work may be treated as professional engineering, while a small portion of the total engineering effort may be rated as technological innovation. Likewise, management may also be viewed as consisting of a large, professional management effort coupled with a far smaller innovation management effort. A small, technology oriented company can get started only by putting emphasis on technological innovation. Small companies that do survive must be strong in technological innovation, a statement which is almost a matter of definition. As these small companies grow, the original innovator must attract professional engineers and managers to form an effective team. From there on, the role of the individual, technological innovator is gradually replaced by managerial innovation with whose collaboration a large group of professionals can produce innovative products of varying technological sophistication. Thus the evolution of a successful, industrial enterprise focuses on the transfer from technological to managerial innovation; managerial innovation embraces both professional management and professional engineering and is interlaced with technological innovation.

In a small firm, one individual may embody all the needed functions, and by tradition he is the technological innovator. The leader of the managerial innovation team in an industrial giant may be the founder, though this is not necessarily the case. Considering the complexity of managerial innovation, the background and training of leadership is not so important as its ability to penetrate the nuances of both the technological innovation team and the professional management team. Conceptually, these two teams may be viewed as entwined like the double helix of a DNA molecule, shown in Figures 4.1 and 4.2.

The first section of Figure 4.1 illustrates the start-up period of a technologically innovative company: It emerges from the strong impulse of a certain, technological endeavor, while the level of professional management is almost nonexistent; the company's growth is marked by the emergence of a well rounded, professional management team, which carries comparable weight in the company vis-à-vis its counterpart, the technological innovation team. This situation is represented by the two

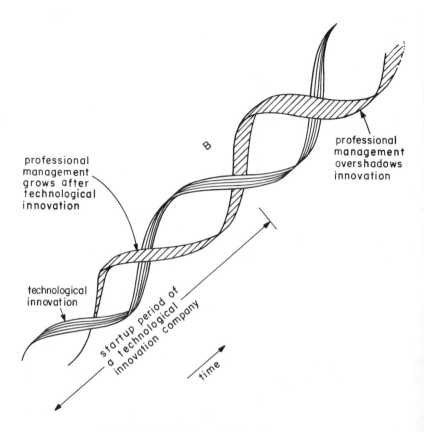

professional
management
grows after
technological
innovation

B

professional
management
overshadows
innovation

technological
innovation

startup period of
a technological
innovation company

time

Figure 4.1. Maturity of a small innovation company.

linked bands which become equal in width at point *B* in Figure 4.1.
Again, as a rule, many more employees are involved in the professional
management group than in the technological innovation group wielding
the same influence.

As time passes and maturity is reached, further growth of the company may assume the pattern illustrated at the right of Figure 4.1. In this
period, the company may in some instances continue to grow with an
expanding, professional management team, while the technological innovation counterpart undergoes continuous decline. As a result, the
growth of the company soon reaches a plateau, and not long thereafter it
too begins to decline.

Figure 4.2 shows the industrial giant under innovative management,
represented by well-balanced, technological innovation and professional

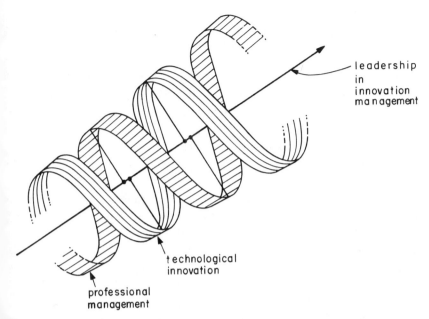

Figure 4.2. Properly balanced technological innovation and professional management.

management teams. At the core of the system is the leadership which provides the strategic mechanism for the harmonious linkage of the two branches so that they may grow toward successful managerial innovation. This double helix conceptualization conveys the idea that a company is a living organ which functions by means of a mechanism comparable to the generic code of DNA.

One possible way of examining the "generic code" hidden in the complexity of innovative management is to treat the professional management and the technological innovation systems as impersonal, information systems in order to identify their intrinsic characteristics. Indeed, the well-being of a company depends heavily on the quality of its leadership. It is commonly assumed that capable leaders are born with inherent capabilities and that many manage on the basis of gut feeling: They don't have time to analyze their thinking processes, and few of their experiences are organized for use as teaching material. This chapter attempts first to analyze technological innovation management and then to verify assumptions through the experiences of a few successful leaders.

2. THE CHARACTERISTICS OF A PROFESSIONAL MANAGEMENT SYSTEM ILLUSTRATED BY FEEDBACK CONTROL THEORY

As an information system, the operation of professional management may be characterized by a closed loop through which monetary information on various activities in the industrial operation flows. In its basic and simplest form, an industrial operation utilizes cash to purchase labor and materials; in due course, these are converted into goods that are sold in the market place to generate earnings, which are then plowed back (feedback) to support the operation, thereby completing one operation cycle. The dynamic behavior of this type of operation is quite similar to that of a feedback control system, which has been studied extensively. Following the practice of feedback control theory, this operation is represented by the block diagram in Figure 4.3, where its basic functions are identified in the forward loop as processing of product and sale and the return path labeled "accounts receivable." The symbol shown to the left of the first block represents a summing point. In this particular situation, this symbol, along with the two plus signs associating the incoming arrows, imply that working capital is the sum of investment plus cash return. This kind of positive feedback is also known as a regenerative system.

This simple, conceptual model provides the basis for examining the dynamic nature of the total operation as a function of the dynamic nature of the individual blocks. For instance, the dynamic characteristics of the processing of product block could be represented by gain and delay time. The "gain" means that the finished product is worth more (or less) in the market than the cost to make it by a "ratio" (not the difference) known as the "gain" factor. But, to realize the market value including the time needed to process the product, there usually exists some delay time. Likewise, some delay may be involved in collecting cash in the feedback branch; therefore, when things work out just right the dynamic behavior of the entire operation may exhibit an exponential growth rate as shown in Figure 4.4. The business will be very attractive when the profit rate it earns is higher than the prevailing interest rate while maintaining a healthy growth.

The concept of control theory was developed to identify the building blocks of physical components that are used to assemble a complete engineering system. Numerous rules and theorems were developed to re

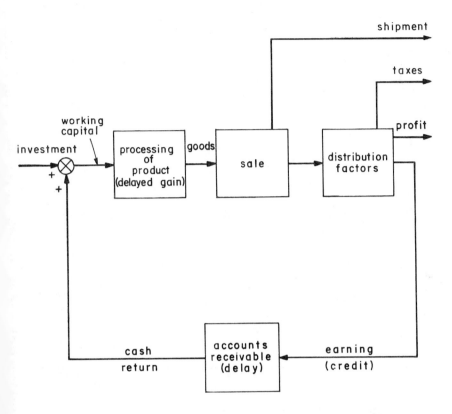

Figure 4.3. Business operation illustrated by a simple feedback block diagram.

late dynamic behavior of the entire system to individual elements through the use of appropriate mathematical models. For physical systems, especially, those elements whose behavior displays fairly linear properties and whose individual performances are "calibratable," the use of control theory to determine behavior of the entire system is very effective. For business operations, the concept of control theory up to the point of using a block diagram to outline the functional relationship is also quite effective, even though using the mathematical model to predict dynamic trend is not quite realistic, as yet.

In physical systems, once dynamic behavior of the components is calibrated, they are usually assumed to remain unchanged, so that the dynamic behavior of the entire system thus established is meaningful. For business management, the behavior of every element changes continu-

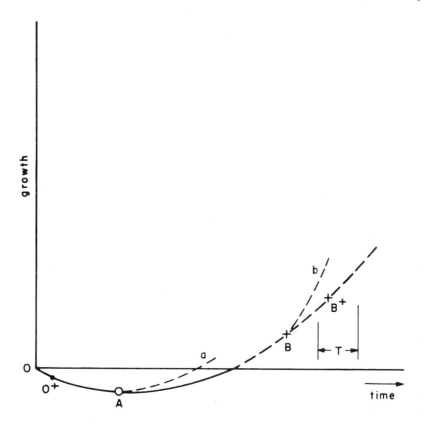

Figure 4.4. Simple growth rate with regenerative feedback.

ously. But if it is also possible to calibrate them continuously (a reliable track record, so to speak), then it is possible in theory to determine the dynamic behavior of the entire system with an appropriate computer mode. At the present time, business management is still handled by the conventional accounting and bookkeeping method aimed at constructing a balance sheet which shows the instantaneous state of various components, while leaving the dynamic behavior of the operation to the intuition of management. Because of this situation, the experiences of the managing team count very heavily; this is even more critical in the growing stage of a new start-up, where a proven "track record" for the new company is simply nonexistent. In this situation, consulting an expert may help in estimating the dynamic behavior of a certain component of the operation which, in turn, can assist in making some meaningful predictions of the dynamic trend of the overall operation.

With the microprocessor swamping the market and many accounting chores, such as inventory control and payroll accounting, already computerized, it should not be too long before a computerized business management system is available, complete with day-to-day, decision making capability. Conceivably, when that does happen, the computer program would have to evolve from control system theory as we now know it. Regardless, prudent managers must set their policy according to future projections, which of course is an extension of the current operation through its dynamic behavior, which can be more easily explained by control theory than a balance sheet. Furthermore, the difference between professional and innovative management lies primarily in the dynamic and statistical nature of the two types of operation. This concept can best be illustrated in terms of control theory. For this reason, an introduction to the dynamic characteristics of various components of the professional management aspect of industrial operations will be discussed in the following section. The relationship between the control system concept and the conventional balance sheet method will then be reviewed in section 4.

3. THE INFORMATION FLOW DIAGRAM OF PROFESSIONAL MANAGEMENT

Figure 4.5 illustrates a conceptual information flow diagram of professional management. It portrays the industrial operation in a similar feedback loop in Figure 4.3, with the forward loop branching out into various expenditure accounts and the cash return loop covering several types of funding sources. In this diagram, heavy lines are used to represent the mainstream of operation, while other thin lines represent various, auxiliary measures introduced to improve the primary mode of operation. The dotted lines represent disturbing effects due to external influences which tend to alter the system behavior. Quite often, the light lines, or auxiliary measures, are introduced to offset damage done by disturbing effects. Aggressive management would strive to improve performance of the operation with voluntary use of these measures.

Four basic categories, or segments, in the mainstream of operation are shown only as an illustration; they are: professionals, workers, materials, and capital equipment. For a stabilized operation, the share of the cost contributed by each segment is supposed to be known. Delay between payment for expenses incurred by each segment and receipt of earnings contributed by each segment is also assumed to be calibratable. Thus it is possible to determine the dynamic behavior of the closed loop system.

The emphasis of the conceptual information flow diagram of professional management is on achieving an optimum earning profile beyond point B or $B+$ of Figure 4.4. For instance, by adapting aggressive, advertising techniques, one may expect the growth curve to be tilted upward, as shown by dotted line b of Figure 4.4. It is to be further noted that while the growth curve in Figure 4.4 shows accumulated assets, the objective of professional management focuses on improving the growth rate (averaged over the characteristic time T) beyond point B when technological innovation is completed, so that the major element of uncertainty in the operation is removed. Skill in reaching point B is considered a function of technological innovation.

Figure 4.5 is a grossly simplified picture of a business operation. Nevertheless, from this diagram one can see that each operating element not only contributes its share of "gain" (This is terminology used in control theory: If the product of all the "gain factors" around the closed loop is greater than unity, then the overall performance will exhibit a positive, exponential function) but also its own time delay. For instance, there may be a particular facet of work, such as an expensive engineering job, which enjoys an apparently high profit margin, but takes a long time to complete, so that, on the whole, it may not be so desirable as another, low profit item which has a fast "turn around" time.

The mainstream of the operation, represented by the heavy line of Figure 4.5, is the essence of the business; in principle, in order to justify its existence, it should be able to generate a profit rate equal to, or better than, the current interest rate, while maintaining or improving its growth rate. A company with an excellent product may grow nicely without additional infusion of funds and is sometimes nicknamed a "boot-strap" operation. New funds may be introduced through various combinations of loan and investment and can help to change the slope of the growth curve, illustrated in Figure 4.4 by two dotted line curves, a and b, initiating from points A and B, respectively.

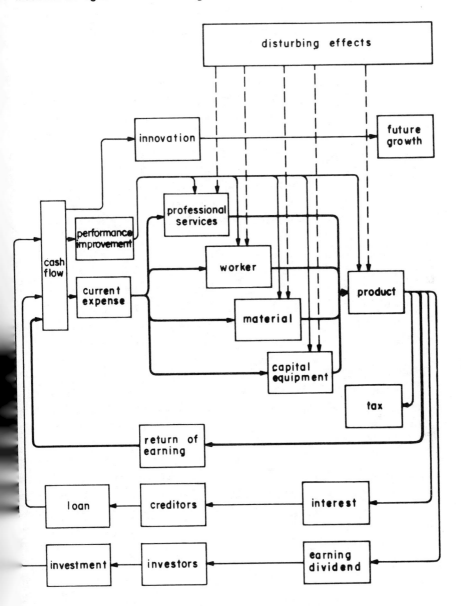

Figure 4.5. Information flow diagram for typical professional management.

Improved performance may be induced through the following three basic approaches:

1. Expand the existing mode of operation by increasing the sales force, enlarging production facilities, hiring more workers, etc; the growth time of this type of operation is relatively easy to predict.
2. Effect "performance improvement" represented by various measures shown in Figure 4.5, including introducing new marketing strategy, styling and model changes in the product, or modernization or automation of the production facility; these kinds of improvements probably take longer to realize than approach 1), however, the result is still quite predictable, and the yield can be even higher.
3. Employ innovation, shown in Figure 4.5 as an open-ended effort aiming at future growth; its effect does not belong to the current business operation. Because of the long time factor (several years) associated with any activities justifying the name of innovation, their outcome should not be accounted for in the feedback loop of current operation.

Generally speaking, only in rare cases would an industrial firm grow as a "bootstrap" operation. Injection of outside funds is unavoidable to maximize growth potential as soon as possible, especially in the current industrial environment, with the exceedingly rapid technology evolution and rate of obsolescence.

With the three major categories of performance for improving modes of industrial operation outlined above and the general growth pattern shown in Figure 4.4, a better summary of external funding can be made in the following manner: Clearly, in Figure 4.4, point B is most advantageous along the path of development of a new business, because it is the best position from which to get a loan, or its equivalent, for carrying out the performance improving approach of items 1) and/or 2), outlined above; point B in Figure 4.1 also corresponds to the maturity of a small innovation company. (Maturity refers to having achieved a balance between innovation and professional management activities.) It should be noted here that the performance improvement approaches of items 1) and 2) are functions of professional management teams; indeed they are the function of a higher caliber of professional management team than those needed to run the routine mainstream of the feedback loop.

At point B of either Figures 4.1 or 4.4, a bank loan can be obtained on easy terms, which means a relatively lower interest rate (such as 1% or even ½% above the prime rate) and no undue collateral attachment. If the loan is used properly with rather straightforward performance improvement approaches and a relatively short time delay (this means innovation does not fit into this criterion, i.e., innovating with borrowed money is not necessarily good strategy), then any gain above the interest owed is pure profit. A littler further along the solid line curve of Figure 4.4, point B becomes $B+$. At that time, if the operation is running smoothly, then additional funds should be injected in order to continuously improve the performance of the operation with the same two approaches, items 1) and 2), listed above.

At this time, the track record of the operation is further solidified. A bank loan is still the first line of outside funding to be used, because, as long as one can maintain a good profit margin, any net gain over the interest rate is a very clean profit, especially since the interest paid is tax deductible. With an excellent track record at point $B+$ and the bank loan reaching a certain uncomfortable upper range, if the potential yield of professional management activities are still not yet exhausted, then additional outside funds should be sought through new investments.

In the above paragraph, two parameters were identified: the comfortable range of bank loans and the potential of professional management; both have subtle implications and involve other supporting parameters. The "comfortable range" of the bank loan refers to the risk factor involved in such a loan: From the bank's viewpoint, there is the risk that the company may not be able to repay the loan, while the company must face the possibility that, in an economic squeeze, the loan will be called in. When a company is doing well, banks tend to be overly eager to make loans, because they think the risk is low; this is the time when a company should guard against borrowing in excess of the comfortable range. For example, the banks themselves may run into financial difficulties and suddenly want to call in their loans, an unlikely event, but one which has been known to occur and to cause serious trouble to some overextended companies.

The second parameter—professional management potential—opens up an entire vista involving marketing strategy, production management, and personnel development; in short, the entire range of skills taught in the regular business school curriculum. These skills are available and intended for use in fully exploring the potential of a newly developed and market tested product. Pushing for solid growth through professional management is an interesting position for the new firm and a challenge to the management team. At this point (or any other), long-range innovation should be kept up through the use of internal resources with the idea that such investments can be written off as a high risk. Funding for long-range innovation, in competition with the fast reutrn on investment offered by a strong, professional management team, may sometimes become unattractive without the support of a strong, innovative leader.

Thus, in the entire history of a company's growth, innovation plays the unchallenged, dominant role during only the initial stage from point 0 to point B of Figure 4.4. To get the process started, the founder/ innovator must usually use funds available to him through personal relationships to initiate the dream by moving from point 0 to point 0+ (Figure 4.4). At 0+, he should have enough evidence to prove to himself that he is indeed involved with something of significance and to convince his potential investor to provide enough capital to aim for point B. The longer he huddles between points 0 and 0+, the better prepared his game plan will be for successfully gaining point B. Once he receives the funding, he must conserve "fuel" in order to be sure of reaching target B. As on a space trajectory, refueling at midpoint A is very difficult, if not impossible, because here the track record is worse than at points 0 or 0+. If additional funds are needed midway, it can mean only that the initial goal of innovation cannot be met. Funding for revitalizing innovation is simply too risky to attract the bank or other sources for a loan, and another investment source must be sought, if development is still promising, but the innovator at this time must deal with great, personal sacrifice.

4. THE OPERATION OF PROFESSIONAL MANAGEMENT

The information flow diagram in Figure 4.5 is a conceptual picture of industrial operation. In practice, professional management uses accounting procedures, balance sheets, and financial reports to measure the performance of the operation. The basic skills for running a business and the appropriate measures to improve its performance are generally described in a very large collection of books and taught in various business schools; it is, therefore, not the objective of this work to dwell on those intricacies. Furthermore, the earlier suggestion that in the near future computer programs may be developed to help the manager make day to day decisions is, basically, a speculation. By and large, most managers do rely on "gut feeling" in decision making, because in this manner most issues can be resolved more rapidly than could be accomplished by detailed, computer processing of the data for optimization.

Despite these qualifications, the information flow diagram can be used to show the framework of professional management in order to set the stage for discussion of innovation management. In essence, professional management operates within a feedback framework (portrayed in Figure 4.5), whereas innovation (shown in this same diagram) is an open-ended addendum which carries no weight when acquiring a bank loan and has little effect on the current operation. Sometimes, innovation is present simply because the company can afford it. As an example, during the fifties and sixties, many prestigious companies rushed into building their own research centers, many of which were established more as showcases in an effort to "keep up with the Jones'," than as functions of a clearly established plan under the guidance of managerial innovation.

As stated earlier, innovation is strictly related to future growth and, except for initial investment to move from point $0+$ to point B in Figure 4.4, innovation funding must be provided by the excess profit earned by the primary business operation after it grows beyond point B in Figure 4.1. Thus, healthy innovation management, as portrayed in Figure 4.2, must begin with healthy professional management as the base to support innovation and reach future goals.

Whether the industrial operation is to be run by gut feeling or, eventually, by computer, the unique feature of its manageability lies in the existence of the feedback path which consists of tangible information. In theory, at any point in the operation, every action is accountable for its expenditures and objectives within the time span (characteristic time T shown in Figure 4.4) which is meaningful to the current operation. It is interesting to note that this characteristic time span is very important to management, so that information sampled at selected intervals is effective for monitoring the operation. For example, the grocery business is a day-to-day operation, while in small instrument manufacturing, the dominant measure of time is about one month; manufacture of larger systems may have to go by the quarter or longer. For all of these operations, annual, fiscal reports represent a fairly smoothed out general trend and, unfortunately, not a long enough time period to cover innovation.

One of the conditions for the maturity of a new business as it is reaching point B of Figure 4.1 or Figure 4.4 is having a professional management team to answer the needs of various segments of the business; In current practice, these needs are monitored by accounting procedures and summarized in the financial report and balance sheet. Samples of a healthy financial report and balance sheet of a small company at the point of maturity are shown in Tables 10-1 and 10-2.

It is interesting to note that the information flow chart in Figure 4.5 is constructed primarily with the pertinent information of the flow rate of various elements. This is equivalent to monitoring the performance of an automobile or other physical plant where the equilibrium of operating conditions corresponds to flow rates of various ingredients or physical parameters, such as air, gasoline, and distances. In the flow diagram of the industrial operation, pertinent information includes material rate, labor rate, sale rate, power consumption rate, interest rate and the profit rate, etc; but, the "measurement" used in the accounting procedure deals with only the accumulated effect of each account and not flow rate.

If the flow rate information of various accounts is needed for the purpose of guiding management, as illustrated in Figure 4.5, it must be obtained by taking the difference between the accumulated amounts of each account at two successive sessions over a specific period. This may appear to be an easy task, but in practice it is a major step which must be approached diligently, in order for a new innovative company to become an efficient and professionally managed organization; on the other hand, it is equally true that it can be over done. For example, once a rigid and efficient accounting system is established for one particular purpose, it may be so rigid and overpoweringly efficient that it acts as an "antibody" to anything new, such as innovation development programs. For this reason, it is important to know the objectives of the accounting system and to use it only to satisfy these objectives.

As stated earlier, an accounting system provides two kinds of information: 1) the accumulated value of each account; and 2) the flow rate of each account. The accumulated value must be determined to get the flow rate which, in turn, is a useful guide in managerial decision. In addition, the accumulated value is need to establish the "net worth" of the company, an assignment of value required by banks to compute the risk factor of a loan. It is also important as a reference in assessing the equity of new stock issues: The Securities Exchange Commission requires two consecutive years's balance sheets and financial statements (certified) to support any public issue of new stock.

Accumulated assets and cash reserves also represent important information that the innovator should watch closely as "fuel" with which to coast from point 0+ to point B along the growth path illustrated in Figure 4.4. To make a valid trajectory analysis, he must constantly balance the projected need according to the rate of consumption against the reserve. For the benefit of innovators or aspiring entrepreneurs who generally lack training in bookkeeping and accounting procedures, it may be interesting to note that one of the greatest innovations in bookkeeping is the double entry procedure. It seems to be the only effective method of keeping track of a constantly changing system, represented by the simple, information flow diagram of Figure 4.5, but involving hundreds or thousands of items as in an actual system.

The logic of the double entry bookkeeping method may be explained by a flow diagram like that of Figure 4.5 which may be considered as the hypothetical model of the actual operating plant. Information storage stations and accounts identified as ledgers are placed at the junctions of the information flow lines. In principle, whenever a transaction is carried out in the actual operating plant, a corresponding piece of document passes through the information flow line of the hypothetical model to monitor the operation of the actual system. Ideally, perhaps with the help of computers, the information should be recorded simultaneously with the actual event. In practice, however, there is always some delay due to the labor involved in making the record. To minimize the chore, the information to be recorded (invoices, time cards, material issue slips, etc.) is allowed to accumulate, so that it may be recorded in bunches. When this occurs, the accounts or ledgers at the two ends of the flow line must be recorded simultaneously, with the receiving end of each station identified as the credit entry and the outgoing end as the debit entry.

In a start-up company, the simple chore of bookkeeping often becomes a problem, and information storage accumulates beyond the point where it can be effective as a monitoring mechanism. While in a healthy, growing firm the actual system runs itself without the need for constantly updating the information flow, it is very dangerous not to do so if the cash flow is in a precarious position.

Bringing a small innovative company to maturity with efficient, professional management is not an easy process because of the high level of professionalism the undertaking requires. This professionalism is quite different from that involved in technological innovation and must be learned by the innovator as his company reaches maturity; however the term "professional" implies a job which is definable and tangible. The task will become more difficult when managerial innovation is needed further along in the company's growth pattern.

5. THE DEVELOPMENT OF MANAGERIAL INNOVATION

The growth patterns of a new innovative firm shown in Figure 4.4 indicates that the activity was initiated by an innovator responsible for using his own resources to advance from ground zero (0) to zero plus (0+), thus proceeding with outside support toward maturity at points B and $B+$. Thereafter, future growth would rely on the skill of the professional management team. Knowing that every product has its own market life, continuous growth of that company would need additional innovations, and the question arises, can a process of innovation development similar to that employed initially be repeated again and again, with many innovations developing simultaneously from ground 0 to 0' and maturing sequentially? The answer is that this can occur when the original innovator is capable of leading the professional management team, while remaining resourceful in leading the growing innovation team. But how could this goal be realized in a general situation where the chief executive is not the technological innovator of every product the company produces? This question must be examined in two parts. The first is the fundamental question of how innovative ideas are nurtured in general, and whether such nurturance should be carried on by independent innovators or members of a large firm. The ingredients for this portion of the study include "perceiving needs" and analyzing various processes for developing skill in technology innovation and will be discussed in the remainder of this work.

The second part of the answer to the above question deals with the environment a large company must create for nurturing innovative ideas. Like trees in the forest, professional management dominates the scene by providing the substance but, at the same time, tends to shade the area from sunlight, block air current, and erode the nutrients (all elements needed by the seedling). How to provide a nursery environment in a dense forest is the challenge to managerial innovation which is to be examined in the remaining sections of this chapter. The discussion will begin by observing some ill effects of the big forest environment on the seedling and will then introduce a logical remedy.

6. ACQUISITION OF TECHNOLOGICAL INNOVATION THROUGH MERGER

One of the industrial phenomena of the 1960s was the emergence of conglomerates of which the Textron Company, led by Royal Little, was the pioneer and undisputed leader. The basic mode of operation in these companies was to begin with a team of management experts and then seek start-ups with innovative capability who lacked professional management skill and marketing know-how. Those were the boom years for new venture business: Capital was in great supply and so were aspiring entrepreneurs; however, all these ingredients were mushrooming in a haphazard way, so that it made good business sense to coordinate the above two elements with professional management expertise. The general formula for acquisition and merger at that time was to retain the original innovator to run his division but to strengthen it with managerial skill, while allowing significant incentive to further motivate the operating team. In this manner, the strength of all ingredients necessary for innovative development was retained, while weaknesses were mutually overcome.

The operation of the conglomerates was in many respects like a venture capital investment house, such as General Durial's American Research, but on a much larger scale. Using the growth curve in Figure 4.4, the venture capital houses specialize in making investments to start-ups at point $0+$, while the conglomerates acquire or merge with companies beyond point B. Thus, the conglomerates deal with "matured" venture business, as previously defined, and therefore should face less risk, but at a stiff price. A great deal of time, skill, and luck was involved in the acquisition process, beginning with the search for companies potentially worth purchasing. The search was then followed by an investigation of the key parameters of each company's innovation, market potential of the product, complimentary nature of the candidate company with respect to that of companies already acquired, personality traits of the principals, strengths and weaknesses of the management team, balance sheet, etc. Finally, at the bargaining table, each side would try to outsmart its counterpart on future projections by claiming the strongest contribution, thereby hoping for more favorable terms in the transaction. This represented a difficult trading process, because the target companies worthy of consideration were in no hurry to sell. Quite often, once the news was out that a particular company was being approached by a certain conglomerate, another one would sneak in putting the target company in a much stronger bidding position. In the end, the deal might be consummated not so much because of a better match of technological, marketing, or managerial issues, but because of the personal charm of a negotiator.

On the whole, in its early phase, a conglomerate tended to concentrate on the acquisition of a few companies. Their professional skill, which was, in theory, to be matched with the skills of new acquisitions to improve their growth potential, was, however, diverted to the task of acquisition. In the later phase, when enough new divisions had been established, a new problem had to be faced concerning application of managerial innovation skills which, unfortunately, is not so straightforward as simply matching a technological innovation team with a professional management team. By and large, the conglomerate concept of matching technological innovation with imported managerial skill came and went along with the fly-by-night venture business of the sixties.

This does not mean that acquisition of technological, innovation capability is not a sound business policy; on the contrary, it is a very effective way for an industrial firm to reinforce certain weak points in its total innovation program. In this respect, the focal point of the operation rests with the managerial innovation policy, which should be in full control of the various, innovative ingredients in planning for the future. In this manner, whether or not each ingredient is developed in-house or acquired in a package deal is a minor problem in comparison with the total plan.

7. THE PHENOMENON OF INEFFECTIVE INNOVATION MANAGEMENT IN A LARGE COMPANY

Some characteristics of ineffective use of funding for new innovation projects in large companies were described in an interesting paper by Jordan J. Baruch.* He observed that characteristically this kind of funding usually exhibits the pattern illustrated in Figure 4.6. Here, significant funding was authorized when the potential level of the project reached a certain critical value A according to the funding practices of the company. Then, with the help of funding, the potential was raised

---

* "Technical and Management Note" by Jordan J. Baruch, EEE Transactions on Engineering Management Vol. EM-21, No. 3, Aug. 1974, pp. 105-107.

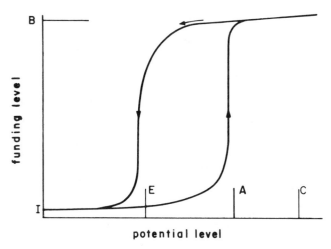

Figure 4.6. Innovation funding practice in large firms.

to level C, which, in turn, induced additional funding. However, if for some reason the project were not successful, and the potential slipped back below level A, the high level of funding would probably be maintained until the potential slid below level E. At that point, a drastic cut in support would occur to dump the program to a low funding level at I. This phenomenon is well known as a hysteresis loop in control theory and usually means some energy is lost if the system is cyclical.

In Baruch's paper, the phenomenon was analyzed by feedback control theory as the result of closed loop system behavior when the behavior of the forward loop is highly nonlinear, as shown in Figure 4.7. This nonlinear, forward loop behavior was identified by Baruch as the "managerial preference curve," which represents the nature of the decision-making process of the executive of the company in response to proposals from the innovation team. In his paper, Baruch explained the relationship between open and closed loop behavior in terms of control theory. However, after this configuration is identified, it is not too hard to see the inherent properties of this kind of relationship, indicating that management tends to ignore incipient, innovative discovery, while becoming overly excited by the promise of superficial gains; this he defined as lack of management sensitivity. In a large company the phenomenon of the hysteresis loop indicates ineffective use of research

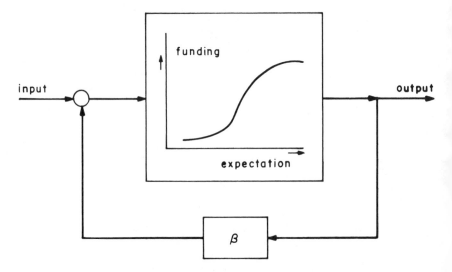

Figure 4.7. Innovation funding practice explained with feedback logic.

funding. For a small, independent venture business, the sudden drop of cash flow (funding level) from point E to point I would precipitate bankruptcy.

In the industrial management feedback diagram in Figure 4.5, the "innovation" branch is shown as an open-ended operation, or an open loop configuration, which seems to contradict the feedback model introduced by Baruch; "innovation" is shown in Figure 4.5 as an open-ended operation because:

1. The feedback system of industrial management operates with a turn-around time of two or three months, so that most of the operational procedures can be established and smoothed out over the conventional fiscal year, in contrast to innovative development, which shows results after several years. Thus, from the point of view of the dynamic characteristics of professional management, innovative development behaves in an open-ended manner. For this reason, in conventional accounting, long-range R&D may be treated as future investment or simply written off each year.

2. The feedback operation in professional management involves tangible information which can be accounted for continuously,

whereas in an innovation development system, the operation is governed by the managerial preference curve, defined by Baruch. This managerial preference curve certainly does not supply tangible information and, as the name implies, is not accountable.

In companies with professional management systems, the innovation development system is indeed very difficult to manage. Because innovation itself is hard to come by, the combination of elusive innovation and equally elusive managerial preference resulting in low probability of success is not hard to imagine. This is especially so when managerial preference comes from a chief executive who is more accustomed to the tempo of professional management and is impatient with the natural trend of innovation development work which cannot be hurried.

8. THE ESSENTIAL ELEMENTS OF INNOVATION DEVELOPMENT

One commonly recognized pattern within the innovation development system involves a "champion" and a "sponsor," with the "champion" representing the innovator and the "sponsor" providing the needed support. Thus, in Baruch's model, exercising "managerial preference" implies the decision made by the sponsor to select a champion. By traditional measure, winning support is the first step toward success for the champion, while the real payoff for the company depends upon successful innovation development. Edward Roberts, in his paper, "Generating Effective Corporate Innovation,"* provided a glimpse of the interplay between people involved in the innovation development process. In his paper, he introduced additional roles in the hierarchy of the innovation development team: The role of the champion was split into two parts: creative scientist and entrepreneur; in addition, a program manager was introduced in the hierarchy under the sponsor. Paralleling the direct chain of command was a "gatekeeper," who provided the team with pertinent information. The role of the sponsor was described by Edward Roberts accordingly:

"The sponsor may in fact be a more experienced, older project manager, or former entrepreneur who now has matured to have a softer touch than when he was first in the organization; as a senior person, he can coach and help subordinates in the organization

* Technology Review, October/November 1977.

and speak on their behalf to top management, allowing things to move forward in an effective, organized fashion.''

The image of the sponsor described by Roberts implies wisdom which, when exercised effectively, could provide the smooth, managerial preference curve characterized in Baruch's feedback model, thereby avoiding drastic, hysteresis loop behavior as the end result of the innovative operation.

The feedback model envisioned by Baruch and the role models portrayed by Roberts each touched on a certain aspect of innovation management; however, the all important criterion—managerial preference—has yet to be classified. Equally important is the interface between sponsor and champion, which must work effectively even if the roles of the individual players satisfy Roberts' description.

In an attempt to understand the mechanism of the innovation management process, the following hypotheses are established at the outset.

9. HYPOTHESIS OF THE IMPORTANT INGREDIENTS OF A MAJOR INNOVATION DEVELOPMENT PROJECT

1. Feedback

   The concept of feedback is very important in a management system. In the previous section, the advantage of having a feedback loop in routine business management was well demonstrated. In that system, the "inherent" feedback loop provides the guideline for running the day to day operation. In innovation management, the feedback must be deliberately created; for example, a pre-assumed target may be adopted to serve as the reference to which future accomplishments of innovation development are to be compared.

   One important aspect of having a target plan is that it allows other branches of the operation to have corresponding matching plans in order to achieve a smooth and efficient overall operation. Equally important is the use of a pre-assumed target plan to define individual responsibilities.

2. Motivation and Reward

   Innovation is basic to human nature. The inner desire to innovate and have the satisfaction of seeing tangible results is as strong as other human desires. To set a person on the road to innovation, there must be a well identified goal, or need, and, along with each positive goal, a set of constraints, such as cost to develop and time to complete. In addition, there must be a person or group of people to whom the innovator relates. The best human environment for stimulating innovation incorporates appreciation from those whose opinions are valued; "peer appreciation" psychology is also deep-rooted in human nature, and appreciation must be reciprocal. The most common situation is expecting praise from one's boss, though he is likewise expecting praise from his subordinates. Good team spirit is usually created by the existence of such mutual bonds of admiration. An able leader not only genuinely enjoys being part of the team but can also observe the team's spirit from a detached position in order to assure that no one in his innovative group is left out.

By comparison, financial or material reward is a rather mechanical device whose essence is fairness and anticipation of difficulties in order to avoid dissatisfaction, which can preoccupy the mind of the innovator and consume a great deal of his creative time. Financial reward should be adjusted infrequently but can include annual pay raises, bonuses, profit sharing, and long-range participation in the growth of the company, reflecting the accumulated contribution and potential of each individual. Giving novel financial or material reward at the appropriate time can stimulate better human relationships. However, this is only frosting on the cake; many able leaders are respected for their "no-frills" attitude.

3. Target Plan

Creating a target plan for innovation is different from creating a plan for business expansion: Innovation has many more unknowns than established, known business. For this reason, in developing an innovation project, the target plan must be reviewed and revised continuously, usually upward and often at several times the original expenditure. Financially, it is important for under-budgeting to be anticipated and prepared for. Even more interesting is the way in which the target plan should be revised to remain effective as a reference for evaluating performance, while keeping in mind the goal of stimulating maximum motivation for innovation. At first glance, this is equivalent to measuring with an elastic yardstick, a clearly illogical approach. On the other hand, if the target is provided as a beacon to guide and inspire, then flexibility of the target is equivalent to providing a new beacon, which is nevertheless continuously in sight; how to establish the new beacon is the challenge of the leadership.

Two important aspects of using a target plan as a reference are reality and flexibility. Setting up a target plan is equivalent to an agreement between the supervisor and his team members. When team members are eager to please, each tends to underestimate unforeseen difficulties and volunteers a higher commitment. If and when the program begins to lag, this commitment induces members to work longer and harder to fulfill the target before frustration sets in. A wise leader is one who knows when to revise the target plan to keep the team eager for the next project.

4. Staging of the Project

The target plan for innovation gradually evolves to maturity when the product developed by the innovation team begins to satisfy the specified criteria necessary to meet the market need. At this juncture, the center of action may be shifting from the innovation team to an implementation team under a "project manager."
According to Roberts:

"The project manager is a still different kind of person: an organized individual, sensitive to the needs of the several, different people he's trying to coordinate, and an effective planner; the latter is especially important, if long lead time, expensive materials, and major support are involved in developing the ideas that he is moving forward in the organization."

Developing an innovation project, as a rule, begins with perceiving a need which is then followed by the initial concept of technological innovation: engineering model development, prototype model development, pilot production model development, etc. All through these stages, there will be different people with different talents and different personalities to provide the drive. In a large company, an expert may be required to move from one project to another to perform his specialty at the appropriate stage of each.

5. The Role of the Sponsor

After analyzing the basic ingredients of an innovation development project, including the team, staging, motivation, target plan, and dynamic nature of the target plan, there remains one elusive element relating to all these ingredients yet to be clarified, which is the function of the sponsor. The best way to clarify this function is to use the information flow diagram presented in the following section.

10. THE INFORMATION FLOW DIAGRAM OF TECHNOLOGICAL INNOVATION AND THE KEY ISSUES IN INNOVATIVE MANAGEMENT

In the last section, several essential elements of innovation management were identified; some were recognized by other authors, some were observed as common sense. The information flow diagram of innovation development shown in Figure 4.8 attempts to tie all those elements together in a logical manner to illustrate further their relationships and the dynamic characteristics of the entire system.

In the center of the diagram, implementation of innovation development is shown; here resides the real effort carried out by creative engineers. As the program progresses, the bulk of the work load may be shifted to production planning headed by a project manager. (Premature formalization of a rigid management structure, including the appointment of a project manager, may be a bad policy.) A "target plan" activated by command of the firm provides the reference to be used to measure the accomplishment of the implementation team. The difference resulting from the comparison provides the motivation and the guideline for action; thus, the main line of innovation development spans the target plan and final accomplishment. Along this path, most of the activities and expenditures occur. The essence of the command is expressed, however, through a set of conceptual, information paths represented by the dotted lines. A conceptual model is developed prior to setting up the initial target plan, and from there on the conceptual model always functions to anticipate market need, which may change as the

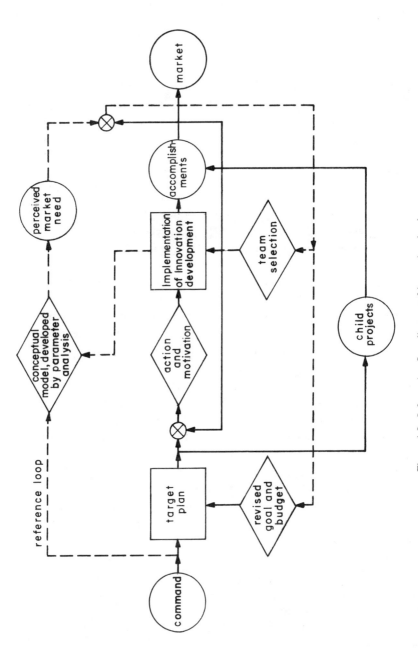

Figure 4.8. Information flow diagram of innovation development.

product is developed, necessitating a new target plan, as well as the shifting around of the operating team.

In formulating the conceptual model, the chief innovator, i.e. the sponsor, must rely on the parameter analysis methodology previously described. Through parameter analysis, the chief innovator would be able to construct the major architecture of the innovation with the implementation team working out details. This conceptual model requires frequent updating by observing various, specific, critical tasks, and, accordingly, conceptual information flow paths overlap in the management chain of command.

The information flow diagram in Figure 4.8 is a "self-adaptive system" according to the terminology of control theory. The dynamic behavior of this system has a built-in capability for regulating expenditure according to step-by-step accomplishment and market trend; this corresponds to Baruch's linear managerial preference function. As an example, one of the cardinal rules of innovation development is to evaluate the dominating parameters of each idea. Many innovations are attractive because of certain novel schemes which represent the improvement of one particular feature over that of the existing product, but as a result of modifications to incorporate the improved feature, some other features will be disturbed and become problems. Skillful evaluation by parameter analysis and judiciously conducted experiments will weed out unworthy parameters without great expenditure.

The key to the success of this self adaptive, innovation development system is the skill of top management in constructing the conceptual model as the basis for continuous reshaping of the target plan from which the implementation team operates and through which a high level of motivation is achieved. The success of every innovative entrepreneur is based on his ability to formulate the conceptual model and constantly revise it to guide his team. Indeed, many gifted innovators can formulate their conceptual models intuitively, and it is believed that the mental process they use focuses on recognition of the dominating parameters and astute selection of the most useful among them.

One noted innovative development in recent history was Dr. Land's approach in developing the SX 70 camera for one step, instant color photography, with the goal of innovation dramatized by the "one step" feature which allows the user to obtain a finished picture by simply aiming at his subject and pushing a button without having to dispose of a negative. But the key, technological objective was to combine the positive and negative films of the color system into one single sheet and the exposing and developing processes in the same dark chamber which is part of the camera. Further refinement included a lightweight and compact unit incorporating a wide focal range from close-up to infinity; on top of that, every mechanism from timing control, to flash control, to film movement was automated. From the standpoint of production, Polaroid had to end its dependency on KODAK and move toward self-sufficiency, a great challenge in itself. Finally, such a camera is useless unless it can be marketed at a price corresponding to consumer appetite and at a significant market volume in line with the new set of production plans.

Land toyed with the idea of a camera like SX 70 for many years. He and his chief chemist tried out many schemes for the multiple layered, photochemical film designed to move the photo sensitive, silver-halide dye and reagent molecules through various diffusions and barriers, so that each ingredient needed in color composition would be transformed in the proper sequence and migrate to the proper layer. As soon as he was sure of the key parameters of the film (but much before the total technology was perfected), he began considering the configuration of the optical system in conjunction with packaging the camera. The development team quickly grew and filled the ninth floor of a building in the Technology Square complex at Cambridge. The team is nicknamed the "ninth floor gang," reminiscent of the "skunk work" of Lockheed under Kelly Johnson, the group responsible for developing the U-2 airplane.

As development progressed, Dr. Clifford Dunken was then engaged as project manager to organize the transition from laboratory specimen to production prototype. Several new facilities, including a completely new film processing plant, film assembly plant, computerized central inventory control, and others were designed and built. The development of the SX 70 and its offshoot, the Pronto, are excellent examples of innovation management corresponding quite well to the innovation management flow chart in Figure 4.8. However, one may wonder whether the success of the SX 70 was due to Dr. Land, regardless of the interpretation of the process. To put it in another way, is the process to be interpreted only as an academic exercise, or does it have some merit as a guide for others? In the following section, this question will be examined through additional case studies.

11. A CASE STUDY OF AN INEFFECTIVE INNOVATION DEVELOPMENT PROJECT

A recent patent interference case between a small company A and a much larger company B reveals a drastic contrast in the approach of the two companies in developing two products with identical, novel features.

The device under contest was a sensing element for a pressure transducer. The novel feature was the configuration for coupling the electrode with the pressure sensing diaphragm in a capacitive type of pressure transduction system. The outward advantage of the invention over conventional schemes was its cost benefit through a reduction in the parts count; the more subtle and significant gain of the new invention was an improved thermal stability.

This particular invention was conceived of in the same year by an engineer in company B and by the two founders of company A, with company B ahead of company A by a few months. At the time, company B was earning 20 million dollars in annual sales, while company A was just getting underway. Both companies manufactured similar types of transducers and viewed this invention as an approach with potential worthy of pursuit. Company A filed a patent in 1973, 11 months after its first sale (U.S. law allows a one year grace period after publication, which includes sale). Due to the simplicity of the application, the patent was quickly issued in 1974.

Company B also filed a patent, but unfortunately for them the filing date was two months later than that of company A, despite the fact that they had conceived of the idea first. Futhermore, due to some complication with their patent application, company B's case took longer to process and, finally, an interference with company A's issued patent was declared by the examiner. The initial ruling of the Board of Examiners upheld company A's patent. The Board's decision against company B was based upon its relatively weak position concerning "reduction to practice," as illustrated in a chart where the various steps in the development work carried out by the two companies are tabulated chronologically. According to this chart, company A had been developing this instrument at a somewhat uniform pace, whereas there existed a gap of three years with no activity by company B in developing the device; This could have constituted a "lack of diligence" on the part of company B.

While this patent interference case is interesting enough from the point of view of patent law interpretation, the records of these two companies in developing their product (made public through this litigation) reveal a vast contrast in the innovation process. Company B represents a well organized, medium sized company, where innovation projects are monitored closely through well documented, monthly reports. Funding for this particular project was appropriated and expenditures reported with the level of funding raised periodically to reach the final figure of several hundred thousand dollars. All this illustrated the existence of a well organized, professional management team, and the end result was indeed a well developed product. However, the four year gap appeared to be the hysteresis loop described in Baruch's paper and caused company B delay in getting the product out, as well as an unfavorable ruling in the patent contest by company A. Close examination of company B's records reveals a great deal of engineering design optimization of the product which was not ready and laborious mathematical modeling of the elementary configuration whose properties could be determined more appropriately by an experimental model. All these superfluous operations were very costly and added little to the final performance of the product. By contrast, company A spent only about one-tenth of company B's development cost to get the product out at a faster pace. Their records showed careful studies of the key parameters of the device and simple, experimental programs to isolate the problem areas for resolution.

A typical management-oriented company usually tends to treat budget review as the focal point, with the dominant concern being the level of visible accomplishment in proportion to capital investment. Without the guidance of a conceptual model developed through experienced parameter analysis and firmly supported by the supervisor, the project engineer, who may be the inventor, is inclined to direct his effort toward showy results, such as fancy equipment, a well organized team, elaborate analytical treatment, and fancy instrumentation layout, etc. This represents "action," and, to those who are familiar with management, action in turn represents tangible progress easy to see and satisfying the subconscious need for visible accomplishment, however short term and superficial. This trend is often found in government funded programs or in large industrial firms that decide to enter new fields.

12. A CASE STUDY OF AN OVER-ZEALOUS SPONSOR

In the last section, comparisons were made between the relative effectiveness of innovation development in a small company and a medium sized one. In that particular case, the small and new company happened to be more effective, by no means a typical phenomenon. On the contrary, too many new ventures meet the ill fate of the hysteresis loop described by Baruch not long after start-up. An example would be a company named (for the purposes of this paper) FTM, which was organized in 1968 for the purpose of developing microfiche for computer information storage. The concept was attractive: Each dot, the size of a few square microns, represented one bit of information, and it was possible to store several million bits of information on a film the size of a pea. It was a time when the price of a semiconductor memory was coming down rapidly to compete with the price of magnetic core at about 1 cent per bit, and yet, the general adaptation of a computerized, data retrieval system, such as inventory control, travel booking, and the like, had just begun to take off. Thus, the need for faster and cheaper memory devices was, and still is very real, indeed. With only a crude "read" and "write" scheme for this microfiche kind of memory, and, at the urging of a certain professional investment promoter, the founder quickly raised several million in venture capital; in this venture, the founder was clearly the champion and the investment promoter the sponsor. An interesting question arises—was the sponsor's interest primarily in seeing the device developed? The answer is yes and no; the professional investor's only objective is to make money, and there are at least two obvious ways to make money through the investment route: the nominal one, which most people assume is the long and tedious technological innovation and entrepreneurial route; the other, and in fact the most tempting at that time, which was to play the new venture investment game.

If a man behind the investment is a bona fide management innovator with appreciation for the key parameters of the innovation, he would give his "champion" the kind of counseling which would stretch available funds over a period properly matched with the normal time span needed to bring a technological innovation to fruition. Better yet, he would maintain a continuous dialogue with his champion to bring out the dominant innovation issues, mostly technological, but within the context of an economic structure. A good management innovator is not necessarily a technologist, but a well prepared parameter analysis can be understood by a nontechnical manager, who can then employ it to guide his champion.

The man behind the investment in FTM had neither the ability nor the inclination for effective, innovative management; his mode of operation emphasized the chance to make a "fast buck" by selling out his holdings during the "second issue" of stock when the price of the new stock could be marketed at several times the price of the first issue (something which occured frequently in the sixties). It is also interesting to note that while the investment financier can sell out during the second stock issue, the principal innovator (i.e. the founder of the company) cannot. With this motivation, the financier would be inclined to do everything possible to influence the founder of the company to make showy commitments. This was indeed what FTM did by, for example, promising a product with unproven performance, indulging in premature building expansion, making excessive expenditures for new equipment, and overhiring untrained employees with no assured product-related function.

When the stock was sold after the first issue, a friend of the founder of FTM asked him how much his stock was worth—the answer was 21 million dollars; this was probably several hundred times his original investment. Unfortunately, FTM was bankrupt about only two years after it went public. After bankruptcy, the legal counsel for FTM commented to an acquaintance at a ski lodge in Europe that he believed that at the time of the first stock issue, they could have gotten twice the capital they did and that extra cash would have seen them through. This comment, coming from a professional, company organizer, suggests one of three possibilities: He was thoroughly familiar with the technical problem and felt additional cash would have put them over the top; he was indulging in wishful thinking about doubling the initial investment; or, worst of all, with twice the capital, they would simply have become twice as extravagant, and the desire to solve the technical problems would have been even further diminished as more time was devoted to spending additional money.

The similar failure of management to push innovative programs, described in the above two cases, may be explained by the "information flow diagram of innovation development" in Figure 4.8. In each case there was the absence of a clearly identified "conceptual model" of how the innovation should be developed in order to generate a corresponding target plan. In the sensor case, company B had the project set up as one of many projects for a company of its size. The supervisor tended to be more conscious of the structure of the management system than of the important issues of the innovation. On the other hand, company A was able to get a grip on the key issues at reduced cost and without rigid program reivew, because the manager understood the key issues of the project and thereby adjusted the target plans continuously.

In the second case (the FTM Company), it was the investment promoter who played the role of sponsor. Clearly, in this case the promoter had no "conceptual model" of the technical problem of the product they were trying to develop; in its place, however, he had a conceptual model of how to boot strap the "apparent growth" of the company in order to attract further investment. Thus, in both cases, the projects suffered, not for lack of sponsorship, but because of the absence of a properly organized, innovation management system. In the following sections, two successfully managed, innovative companies are examined against the information flow diagram of Figure 3.8 to establish the validity of the concept.

13. CASE STUDY OF TECHNOLOGICAL INNOVATION MANAGEMENT: ANALOG DEVICE INC.

Analog Device is a company that manufactures signal conditioners and computer interface; as of Spring, 1978, it employs about 1400 people with 50 million dollars in annual business. The company was organized in 1965 and has enjoyed steady growth ever since. With an electrical engineering degree from MIT, Ray Stata was one of the co-founders, beginning as Vice President in charge of marketing and becoming President and Chief Executive in 1974. According to Stata, even though Analog Device was formed as a technology based company, the early emphasis was on marketing, to the extent that some engineers felt they were second class citizens; this may reflect the fact that Ray was a strong organizer in the company and carried great influence in the marketing area. Now he is in total control, with employees housed in several plants in both the U.S. and Europe, and his feelings about the need for innovation have increased steadily over the last few years.

According to Stata, each company has its own management style, a word which represents a very subtle description of how the channel of command flows through the system or, rather, defines the chemistry of human relationships. On the surface, Analog Device is no different from other companies, in that the chain of command begins with top management, where business objectives and projected market area are determined.

Stata referred to "climate" to describe the environment Analog seeks to cultivate. One element of this climate is the rule against punishing anyone for making mistakes. Mr. Stata explained that progress is often made through experience gained from making mistakes; in other words, making mistakes in innovation is a normal way of life. In fact, it is sometimes difficult to determine what represents a mistake until the entire project is completed, because what may initially be considered an error is later regarded as the most profitable avenue of approach. While it is the policy of Analog Device Inc. not to punish anyone for making mistakes, a control system was set up to minimize the number of mistakes which do occur, and it involves the following: 1) Define the range of decision each level can exercise; for example, a project manager may have a certain amount of money allocated to try out new ideas without having to obtain his superior's approval. 2) Define the business and product line from the top; this means that even though each project manager can exercise his prerogative in new ventures, these innovations must come under a predetermined product line or business area. 3) Dismiss personnel only with approval from the top; this policy is probably quite similar to that of many companies which, after all, invest considerable capital to develop their employees, particularly the specialists of the innovation group. All innovators tend to have different (and often contradictory) opinions concerning technical matters, and these opinions should be encouraged in an innovative environment. Friction and animosities, as the result of contradictions, should be avoided whenever possible, but in any event opinions should not be withheld for fear of job security.

In the area of incentives, Mr. Stata put emphasis on teamwork within groups. It is often difficult to distinguish individual contributions in a team effort, and, if a company adopts the practice of giving credit to individuals, the desire for personal reward might outweigh group accomplishment; above all, team spirit must be developed to enhance peer appreciation. The ultimate achievement in team spirit is to have all members on the team share in the satisfaction of helping each other make a contribution. As the head of a fast growing company, Stata placed emphasis on incorporating new talents—good people attract good people. Analog Device Inc. adapted an apprentice system by assigning a few new engineers to be trained under a top man, and, to further encourage innovation, Analog has adopted a dual ladder system to give innovators special recognition. This system is stressed by Stata as a means of correcting the emphasis placed on marketing in the earlier days of the company. Through the dual ladder system, technical personnel within the corporate structure and some senior engineers who reside in the profit centers gain status equal to that of the marketing executives who dominated the company in the paast.

Analog's long-range business strategy focuses on shifting gradually from manufacturing components to subsystems and then to sensor-based information systems. The new product programs are reviewed frequently by the strategic planning committee composed of key personnel in the company. It is apparently their reasoning that components serve a limited market unless coupled with systems; the sensor-based systems actually include all components, in addition to all interconnecting devices. To achieve its goal, the company will develop certain technologies and acquire existing technologies from other companies; for example, it recently acquired a microsensor corporation in Holliston, Massachusetts and arranged a merger (described earlier) when Analog expanded into integrated circuit technology.

The purpose of the interview with Ray Stata was not to document another "success story" but, rather, to fit his experience into the technological innovation management information flow diagram in Figure 4.8 and systematically identify the various key parameters and their couplings in innovation management. Since Analog Device Inc. is a centrally controlled, multi-product company, the diagram in Figure 4.8 should be viewed as consisting of several, parallel branches of target plans, all monitored by one conceptual model—the "business strategy" of Analog Devices. The function of the conceptual model is to look beyond the feasibility of certain key, technological parameters and match these parameters with perceived market need. The conceptual model should always direct the implementation branches and provide meaningful guidance.

Analog Device Inc.'s ten-year business strategy for expanding their currently predominant component product line into subsystems and, eventually, sensor-based systems represents confidence in the following key issues:

- Sensors represent the equivalence of human nerve endings.

As explained in Chapter I, nerve endings appear to be the last underdeveloped area in extending human biological functions. Sensor-based systems represent the coupling of various kinds of sensors with microprocessors in some individual, functioning systems for various applications, such as medicine, production control, mining, etc.

Market competition will be strong, but sensors are a new and growing field where Ray Stata's skill in marketing and his ability to identify customer's need will give Analog Device a decisive advantage.

- The technology involved in sensor-based systems and analog-digital, interface components is close enough so that the expertise involved in making a parameter analysis of all the products to be developed in the foreseeable future is well within the grasp of the present management team.

In carrying out the target plan and implementation of technological innovation, Stata made it clear that he does not consider himself an inventor, which means that he does not relish going into the laboratory to test some of his own pet ideas. In many respects, this is an advantage, especially for a multi-product company, because, without identifying with any particular product, he can exercise technical judgment on all products equally on an objective or marketing oriented basis. The following two examples may be used to illustrate his skill in conceptual modeling and technological implementation.

- Analog Device started business with an operational amplifier which was an outgrowth of the Analog computer. At that time George Philbrick (the pioneer in that field) owned a company which dominated the domestic market, but he was an inventor and designed his product according to his own, inventive mind. Ray Stata took the basic technology to the user—especially the European users who, at that time, were hungry for U.S. know how but lacked close, technological coupling. In this manner Analog Device's operational amplifiers were tailor-made for the customer and were, therefore, integrated into O.E.M. items.

- One of the major decisions Stata made was to have an in-house integrated circuit manufacturing capability when Analog Device had only 7 million dollars in annual sales; the need for in-house integrated circuit manufacturing capability will be discussed in the next section. Of particular interest here is that at that time Stata and his management team had no technical experience in integrated circuit manufacturing. For this reason, their skill in doing parameter analysis and formulating a conceptual model to guide the target plan for that venture was very limited. This meant that if they tried to hire "experts" to open up new factories for that task, the risk would be much too great for the size of their operation. Their approach was to engage three experts in a new, joint venture with Analog Device Inc., retaining a certain buy-back right. In this manner, the technological responsibility was shifted to the new management team through their active participation.

Ray Stata's general philosophy of innovation management consists of two levels of activity: a strategic management level and an operational management level. Conceptually, strategic management is represented by the central axis of the "double helix" in Figure 4.2. It is the function of strategic management to supervise the professional management team represented in Figure 4.5; it is also the function of strategic management to provide the conceptual model and need perception illustrated in Figure 4.8.

14. CASE STUDY OF TECHNOLOGICAL INNOVATION MANAGEMENT: WANG LABORATORIES

Wang Lab is a company that specializes in minicomputers and word processors. They employ about 4500 individuals and in 1977 reached a level of about 160 million dollars in annual sales. The company was founded by Dr. An Wang (famous for inventing magnetic core memory) in 1951; from that time until the present, Wang has always been the principal innovator, as well as owner and chief executive. The first phase of Wang's business operation, from the 1950s to the early 1960s, focused on various, innovative products.

The growth of the company during the fifties and sixties was steady but not sensational; Wang Lab began to "take off" in the early sixties when the Wang calculator was introduced. In the annals of computer science, Wang was probably the first to introduce the electronic desk calculator at a time when many large companies were involved in developing time-sharing (terminals branching from a large, centralized computer to broaden its utilization) or building minicomputers for special applications, as in the case of Digital Equipment Co. Inc. Thus, the decision to tackle the unique market at the low end of the computer spectrum was an innovative move. Matched with his market strategy was his innovative use of the logarithmic function as a building block for performing multiplications. This simplified the computer program considerably and made the Wang calculator a viable product in the market place at the time.

Between 1960 and 1970, the computer industry was highly competitive, with many companies failing at various stages of their growth. Wang's continued success (and the fact that he retains a major share of stock in his company) is probably unique among his peers. Aside from many shrewd real estate deals, public offerings, utilization of bank loans, etc., Wang's major success may be attributed to his vision of the future of his business as defined in Figure 4.8 by the conceptual model of a product's technological features and the market need.

While success is the result of ongoing development of one's business, a few critical decision points exist to mark each milestone of growth. In the history of Wang Labs, three such points may be identified:

1. The decision to develop the desk calculator.
2. The decision to move from desk calculators upward to minicomputers.
3. The decision to move from minicomputers laterally to word processors.

With each new direction, Wang wanted to make sure that the new market niche could be filled by his team's technological innovation capability vis-à-vis the competition; that his marketing team could handle product promotion without too much adaptation; and that the market had sufficient stability. For example, in moving upward from calculator to minicomputer, he resisted the trend to rush the pocket calculator bankwagon, which also represented commitment to a more difficult technical challenge, because the move into the pocket calculator market from a desk calculator market base required only product repackaging development of a new minicomputer was a much more elaborate task In retrospect, it is not hard to see that Wang's was indeed the proper move. By his own account, he regarded the pocket calculator as a con sumer product, subject to rapid styling change (not his type of business), and, furthermore, he probably enjoyed the technological challenge of the minicomputer. However, at such a crossroad, it is difficult for most people to identify all the parameters and to make a choice with successful results.

Wang's interest in the word processing system began in 1968. A word processing system is a glorified typewriter with intelligence, and, with proper software, it greatly enhances human effort in editing, filing, routine documentation, etc. Moving from a minicomputer to a word processing system is a side step in technology; on the other hand, making the transition from pocket calculator to word processor would have been almost impossible had Wang selected this as his earlier move. From the market standpoint, word processing systems are not very different from minicomputers, and often the same company that buys Wang's minicomputer can use several word processors. The same selling and servicing techniques also serve the same customers. The market volume of the word processor appears, however, to be somewhat larger than that of the minicomputer.

Wang's attitude toward technological development is a simple and prudent one: Take small steps and try all possible routes. While Stata makes it a company rule not to punish anyone for making mistakes in innovation, Wang goes a little further by encouraging mistakes. In taking small steps, mistakes can be considered purposefully introduced perturbation for finding the right direction for the larger strides which follow the perturbation test. In carrying out the perturbation test for conducting innovation implementation, guidance is provided by the conceptual model and the updated market need, as shown in the "reference loop" in Figure 4.8. While Wang identified "making mistakes" as a necessary process in innovation development, he cautioned that ultimate success is determined by the total effort required to reach the goal. Thus, a perturbation which deviates from a guided course is a necessary testing process for minimizing major error, even though that deviation is, in effect, a mistake. On the other hand, many companies conduct innovation development without a properly conceived conceptual model and updated analysis of market need for reference; this generally happens either because the "sponsor" who is responsible for that task is ill-prepared or guided by incorrect motivation—errors thus incurred would have to be called blunders, not mistakes.

During the interview with Wang, the concepts underlying information flow diagram for management (Figure 4.5) and for innovation (Figure 4.8) were discussed. The diagrams facilitated discussion because each of Wang's thoughts could be quickly identified in the diagram; he did indicate, however, that in his operation professional and innovation management are not as exclusive as the two diagrams indicated. He pointed out that in the computer business, changes occur at such a rapid pace that the technological innovation cycle would have to move so fast that it would become indistinguishable from the charactertistic time allocated for professional management. Another major reason for merging the two diagrams in his operation is that Wang is both chief executive and chief innovator and, as such, participates equally in marketing and technology. For example, in his organization, there is no marketing division, only the sales division, with marketing being done by the development center which he supervises closely.

Wang considers his company to be sales oriented and cites the fact that out of 4400 employees, 2600 are in sales and service, compared to 1100 in production, 400 in development, and 300 in administration. His recent sales innovation is T.V. advertising, which is new to this kind of business and represents a 900,000 dollar experiment. Wang's estimate of the cost of introducing a new product is that for every dollar spent in development, there will be 10 dollars invested for production and 100 dollars for sales. It is very interesting to note that while Wang started as a technological innovator (he had only one salesman out of 30 employees in the early days), he now takes pride in sales, while Stata, who started as marketing chief, is now a champion of innovation. The reasons underlying these shifts in emphasis are clear: In Stata's company, the focus over the next ten years will be on generating a diversified product line which will penetrate a new and expanding technological area; to accomplish his goal, he must rely upon innovation. Wang Laboratories, on the other hand, must deal with the highly competitive minicomputer and word processor markets. While 40% of Wang's business (which is growing at a rapid pace) is represented by the word processor, in 1978 it enjoyed only 5% of a market dominated by I.B.M. at 85%, with RCA, Burroughs, and others sharing the remaining 10%. Then, recently, Exxon Enterprises announced its entry into this lucrative field. Wang Laboratories is facing a challenge from competition, and there is no question that in his research lab, Wang is trying out many innovative schemes, but, for the immediate future, the battle is to be fought in the market place.

5
THE PERCEPTION OF
A NEED

1. INTRODUCTION

In 1968, Dr. Wang of Wang Laboratories, Inc. decided that, with his company's well developed minicomputer market and its substantial technological base, entering the word processing market (at that time in its infancy) would require a relatively easy adaptation. Wang further sensed that word processing represented a specific application of the minicomputer to word manipulation instead of numbers and, except for the outward objectives, much of the basic techniques (such as, sub-routine, soft ware, data storage and data retrieval) are quite similar; fur-thermore, for every minicomputer needed to handle numbers, there could be a market for five word process.

The above scenario is probably a very realistic description of the first step in the conception of a new product by the top manager of a well es-ablished firm, a process which involves the following considerations:

- identifying an attractive and enduring market with which one is fa-miliar and within which there is in-house experience to interface with the user;
- having confidence that the product can be developed innovatively to surpass that of the competitor; in particular, emphasis should be placed on "uniqueness" to carve out a market niche;
- making a realistic appraisal of the company's resources and capa-bility to complete innovation, development, and marketing phases.

Thus, perception of a need is not a technical solution but, rather, recog-ition of a combination of a few key parameters covering the entire ode of operation to bring a product to the market.

What An Wang, Ray Stata, or any experienced innovator visualizes from ground zero, down the road of product development is a diagram such as that in Figure 5.1. This diagram is, in effect, an expanded version of Figure 4.4 in Chapter 4 and shows the various, major ingredients involved in the innovation development phase of a product growth curve. A few important elements of this diagram are:

total innovation development cost
production preparation cost
sales cost
earnings

Break-even point: the earning rate is zero.
Zero retained earning: all investments are recovered.
O: the starting point of technological innovation.
B: the starting point of the operation of professional management.
T: characteristic time of the product.

In the last chapter, the task of improving earnings of the product beyond maturity at point B was given to the professional management team. Their function was illustrated by the feedback control logic of Figure 4.3 of Chapter 4, operating with a product characteristic time T.

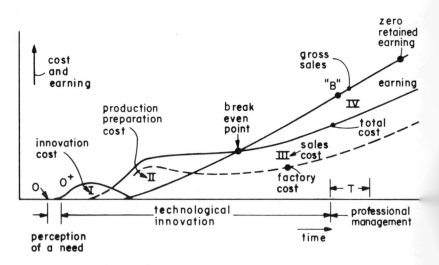

Fig. 5.1. The development of a new product.

The task of developing a new product was identified as technological innovation, and the appropriate way of managing technological innovation in a firm among other professional management teams was described in Figure 4.8 of Chapter 4. Finally, as shown in Figure 5.1, the birth of an innovation project between time 0 to 0^+ is identified as perception of a need: This is the vital moment when the reference loop begins to form, including the conceptual model, and examination of the need, as shown in Figure 4.8 of Chapter 4.

In this chapter, evaluation of this critical moment will be examined with reference to a few real cases and the summarization for establishing the various key parameters involved in this activity.

2. THE CONCEPTION OF AN INNOVATION

When the history of a new product is traced to its origin, one may find that it began as "perception of a need" by the innovator. This "perception of a need" is not a wild dream or casual observation which usually vanishes in the next moment; it is a deeper reflection, beginning with sizing up various aspects of the need in the innovator's mind and the conviction that a technical solution can and must be found to satisfy that need. Guided by confidence, effort will be directed to search for the solution, and, thus, the innovation process begins.

Confidence that a solution exists derives largely from the innovator's earlier experience with similar problems. If he is skillful and resourceful, the exciting process of mapping out the general skeleton of the technical solution which looks most promising may not take long; this may, therefore, be considered the moment of conception in innovation, the point at which the union of a perceived need with a plausible technical solution takes place. Great innovators usually find solutions themselves, but it is not uncommon that perception of the need and identification of the technical solution are accomplished by different parties. If a patent is generated under such circumstances, the inventor's title goes to the one who conceived of the technical solution and not to the one who perceived the need, as defined by the U.S. patent law and by most patent laws in other nations. As an analogy, the inventor is equivalent to the "father" of the innovation, while he who perceived the need plays the role of "mother." As a rule, after a new product is on the market. The field is wide open to other potential inventors to try to do better. Quite often, a skillful inventor, overhearing the discussion of a need, may come up with a technical solution and file for a patent with no concern for the source of his information; conversely, however, one cannot knowingly file a patent on a technical solution conceived of by another person, regardless of who perceived the need. Thus, conception of an innovation is the result of the union of a perceived need with a technical solution, but only the latter is protected by law, giving the inventor a limited period of monopoly. For society, it is certainly logical not to give anyone a monopoly on its needs—even imaginary ones; however, in the total process of innovation; "perception of a need" is the distinctive, first step and in many situations the most important one.

3. THE ESSENCE OF PERCEIVING A "NEED"

Despite the patent law's apparent partiality toward the innovator who conceives of the technical solution over the individual who perceives the need, it is, however, the latter who enjoys a better chance of success, because an invention which is initiated by perception of a real need is usually close to the pulse of the market place where the merit of the new product will be put to the test. On the other hand, there are many self-proclaimed inventors who tend to invent for invention's sake; they are preoccupied with the intricacy of elegant solutions, while the needs they perceive tend to be based only on their own assumptions. Thus, such an inventor may get a patent, even a working model, only to find that in reality, the market is not there. As stated earlier, an experienced innovator would try to get a clear grasp of how a need is defined when he sees it and would be guided by confidence derived from his insight into similar problems to reach a solution satisfying that well-defined need. Quite often, he may find the solution himself and become the inventor, but, through appropriate arrangements, he can also hire or contract with someone else to do the inventing. Either way, the result of his endeavor would be the same, i.e., a new, marketable product. The protection of his rights, including the right to a limited period of monopoly, can usually be included in the contract when he engages other people to do the inventing. As we will see in the following sections, as an innovator becomes more and more involved in the entrepreneurial role in industry, he must delegate more and more of the inventor's role to his subordinates, while, as an innovative entrepreneur, he usually remains the pacesetter of the company by being alert in perceiving needs.

Man is born with competitive instinct, and his needs are quite often a reflection of his discontent with the existing way of doing things: Edison invented the automatic, telegraphic message reporting scheme when he was quite young, because he needed that device to free himself to do other things; Dr. Edwin Land perceived the need for instant photography when he visited Yosemite with his daughter and wanted to capture the image of his child set against the grandiose, natural background. While such inclinations are frequent, they do not constitute well-defined, perceived needs unless supported by the realistic belief that the solution exists within the state of the art and by the determination to find that solution. There are many people who can quickly identify the voids or deficiencies in their environment but lack the urge to carry these observations one step further in order to define and analyze the need and mold it into the embryo of a new product. Since invention is a well-defined legal process and patents are often recognized as tangible property, the art of inventing usually receives far more attention than the process which led to the invention. For instance, when the deeds of a successful innovator are recounted, it is common practice to dramatize the beginning with the occurrence of an "inspiration," such as the Yosemite episode described in the last paragraph.

The emotion that Dr. Land experienced at Yosemite is not uncommon on occasions, such as children's birthday parties or weddings, but often the desire for an instant playback of a passing image only constitutes a regret at having no solution and not perception of a need coupled with belief that the solution to that need could be realized. What is lacking in most people is a spirit of inquiry into the technical background associated with that kind of need in order to be prepared to satisfy it. For instance, many people are familiar with darkroom techniques. After mastering the developer—acid—hypo sequence and the thermometer, timer, and dryer routine, few venture into the basic photochemical action of the film and ponder why the process has to be so complicated. When told that it is the unexposed portion of the film that has to be washed away leaving the exposed portion to turn into the black image of the negative, few extend this notion to realize that the washed away portion is indeed the corresponding, positive image. Thus, by using a "blotting paper" to pick up the portion to be washed away directly from the film, one would have a positive image on the paper, according to Land. It is, therefore, the understanding of existing technology gained through this spirit of inquiry which substantiates a perceived need.

The human being is constantly bombarded by all kinds of stimulants, most of which stay in the mind briefly, and only those which match some latent thought result in action. Thus, perceiving an "applicable need" is a cultivated behavior and not just simple instinct. In the following sections are several examples which illustrate how "applicable needs" are perceived under various conditions.

4. THE NEED PERCEIVED BY THE "USER" OF AN INDUSTRIAL PRODUCT CHAIN

Von Hippel studied the statistics of innovation among scientific instrument makers and the manufacturers of solid state components and concluded that it is the "user" of the instruments or the tools who brought about most of the innovation. Each of these roles represents a link in the industrial product "chain," which means that the users of the product are themselves producers of other products and hence should be well attuned to the spirit of inquiry and logical thought. As the user, he has more field experience related to the "need" for a particular piece of equipment and, for that reason, is more apt to come up with innovations to improve that product and thereby satisfy his own need. A new product will then evolve when the solution to his need can be shared by other users. This is common practice in a variety of industries where the research engineer must devise his own test system and the production engineer his own tooling. In principle, it is more efficient to check around to see if a particular type of tooling is not already being developed by some other firm which has more experience, but in industry, we do go through a lot of the "reinventing of the wheel," which is inefficient and results from ignorance. However, once in awhile users must work out their own needs or join with the best manufacturers of a certain device to push the frontier further in order to satisfy a newly established need. For example, in the design of the SX 70 film pack, Polaroid wanted to have a flat battery to go with each pack. This battery had to be very thin, while still capable of supplying electrical current to drive the camera motor, all at a low production cost. After failing to find a manufacturer of such a battery, Polaroid finally adopted a joint venture development program.

Among the few excellent examples cited by von Hippel, the following are of particular interest:

> As an example of such a 'user dominated' innovation process, consider the innovation history of 'wire wrapping': Wire wrapping is a means of making a gastight, reliable

electrical connection between a wire and a terminal without soldering. It has great advantages over soldering in speed and also allows one of design very dense arrays of terminals without fear that workers, in the process of making a solder connection to one terminal, will inadvertently damage adjacent connections with the heat from their soldering equipment.

Wire wrapping was developed at Bell Labs to provide a means of making electrical connections to a new relay (also being designed at the Lab at that time for use in the Bell Telephone system). The basic wire wrapping process requires a hand tool that winds the exposed end of a wire to be connected tightly onto a terminal of novel design. This hand tool was also designed at Bell Labs, and the entire wire wrap system then went to Western Electric for implementation. The Make/Buy committee of Western Electric decided to have the hand tool portion of the system made by an outside supplier and put it out for bid. Keller Tool of Grand Haven, Michigan, a company which had an excellent reputation for manufacturing rotary hand tools, such as powered screwdrivers, and which supplied such tools to Western Electric, won the bid. Western Electric gave Keller a complete set of drawings for the tool. Keller suggested design changes that, while preserving the tool's basic design and operating principles, would, in Keller's opinion, make the tool easier to manufacture and use. Western Electric agreed to the changes and, in 1953, a purchase order was negotiated. Keller realized that some of its other customers for electronic assembly tools would have a use for wire wrap and so requested and obtained a license from Western Electric which would allow sale of the tools on the open market.

Wire wrapping is now commonly used in circuit development laboratories because it is faster than a soldering iron and casues fewer short-circuits between terminals in congested areas, such as behind the printed circuit board holding "dual-in-line" integrated circuits. Not long before reading von Hippel's paper, the author had occasion to briefly examine such a hand tool and marvel at its rather clever construction. Now, with knowledge of its background, a short parameter analysis illustrates certain subtle features of the innovation process.

• There are many ways to connect a wire to a terminal: some use solder; some clamps and screws; some crimping tools, along with special sleevings. For electrical utility application, this can be a major undertaking, because, while speed of application is important, reliability of the joint against corrosion is even more important. It is quite likely that after the relay designer at Bell Labs finished his work, the assignment was then passed on to a connector designer. After the connector designer finished his job, then Keller Company was called in by Western Electric (Bell Lab's counterpart in production of the Bell System) to have the model of the tool redesigned for production. In this sequence, it is not the circuit board solder man or the relay designer or Bell Labs who invented the tool from the user's vantage point, but the Bell System, as a responsible, utility manufacturer, who, through years of experience, realized the need for economy and reliability in producing their equipment and established a set of criteria for their product. Accordingly, that set of criteria defined the need.

• Wire wrapping is really an ingenious scheme. One way to appreciate it is to realize that compared to the other methods (screw, crimp and soldering) cited earlier, this is the only one which does not require additional parts, aside from the terminal and the wire, to accomplish the coupling (more sophisticated methods may involve laser beams welding, ultrasonic bonding, pressure-heat bonding, etc., but these usually require more exotic equipment and are limited to very thin wires).

Why then was it not used before? The problem is probably that wrapping a wire over a post to make a tight joint is not a straightforward process, because, after wrapping, the wire tends to spring back and lose its grip. The wire wrap tool overcomes this problem by wrapping the wire under high tension so that the bending-mode springback is minimized (see Figure 5.2). Finally, the post is made into the shape of a square rod which allows the wire to dig into the corner adding to the grip. Thus, out of the approximately eight turns of a complete wrapping, the two ends may lose some of their grip, but the center section will maintain the original tightness.

• The combined use of the elastic and plastic properties of material to achieve coupling between two parts is an innovative process worthy of note and can be generalized for other applications. Most people know only by common sense the simple schemes involving either elastic deformation, such as spring clips and screws, or plastic deformation, such as crimping and the ordinary twisting o wires. The combined use of plastic and elastic deformation can suggest many interesting applications. As a rule, patent law doe not protect general principles; however, one can usually apply thi kind of general principle to each specific application with specific configuration and get a patent. Thus familiarity with thi kind of general principle is very helpful to an innovator.

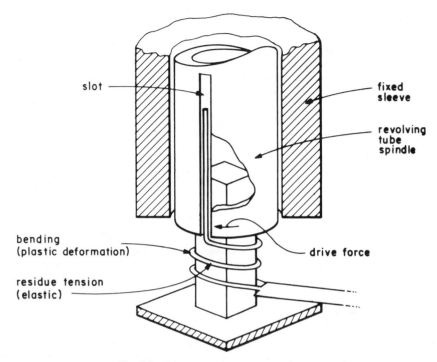

Fig. 5.2. Operational principle of wire wrapper.

1. Insert the stripped wire inside the slot of the revolving spindle.
2. Insert the terminal post inside to center of the spindle.
3. Wrap the wire by gripping the handle of the tool to turn the spindle through a gear train.
4. Tight bond between the wire and the terminal post is achieved by the residue tension in the coil.

5. THE MANUFACTURER INITIATED INNOVATION PROCESS

Von Hippel, in advancing his thesis that innovation is generated largely by the user, also cited a few cases which speak for the manufacturers; one such case is quoted here:*

Case Outline

"A major improvement innovation: well-regulated, high-voltage power supplies for transmission electron microscopes (manufacturer-dominated innovation process). The first electron microscope and the first few pre-commercial replications used batteries connected in series to supply the high voltages they required. The major inconvenience associated with this solution can be readily imagined by the reader when we note that voltages on the order of 80,000 volts were required and that nearly 40,000 single cell batteries must be connected in series to provide this. A visitor to the laboratory of Marton, an early and outstanding experimenter in electron microscopy, recalls an entire room filled with batteries on floor to ceiling racks with a full-time technician employed to maintain them. An elaborate safety interlock system was in operation to insure that no one would walk in and touch something electrically live. Hanging in the air was the strong stench of the sulfuric acid contents of the batteries; clearly, not a happy solution to the high-voltage problem.

The first commercial electron microscope, built by Siemens of Germany in 1939, substituted a 'power supply' for the batteries but could not make its output voltage so constant as it had been through use of batteries. This was a major problem, because high stability in the high-voltage supply was a well-known prerequisite for achieving high resolution with an electron microscope. When RCA decided to build an electron microscope, RCA electrical engineer Jack Vance undertook to build a highly stable power supply and by several inventive means achieved a stability almost good enough to eliminate voltage stability as a constraint on high-resolution microscope performance. This innovative power supply was commercialized in 1941 in RCA's first production microscope.''

The above example clearly illustrates the need for a power source other than a roomful of primary batteries to match an elegantly conceived electron microscope. The design of an 80,000 volt d.c. power supply was doubtless a challenge in 1941, as illustrated by Siemens' failure, and it speaks well for RCA that they succeeded. But this contest between the two giant manufacturers does not mean that all manufacturers need be innovators. If the power supply were not developed and one such microscope were installed in an institution, such as M.I.T. or Harvard or Princeton, with a roomful of batteries, then the user might have jumped the gun in developing the power supply because the need was so clearly established. The basic principle for a stable power supply would be the use of "feedback." Indeed, in 1941 the feedback concept was barely on the horizon. The feedback principle allows the use of a reference voltage to regulate the voltage of the power supply. The reference is free from the burden of supplying the power, while the power supply is free from the duty of maintaining the steady state stability; both elements were available in 1941. For instance, a handful of tiny

*The Dominant Role of Users in the Scientific Instrument Innovation Process.

standard cells (better voltage references are available today) with or without a precision resistor bridge could provide the needed reference and achieve a system performance much better than the open loop dry cell system. Siemens' failure could have been in achieving dynamic stability, which in 1941 was a major task.

6. MOVING FROM "NEED" TO FINISHED PRODUCT WITH PROFESSIONAL INVENTORS

A professional inventor is one who derives his principal income from royalties, taking risks on his own and does not attempt to be a manufacturer. Without his own manufacturing plant, he is free from all the burdens and headaches of an entrepreneur, such as marketing, production, meeting the payroll, etc. In this manner, he can devote most of his time to doing what he enjoys most—inventing. Looking at it in another way, he is avoiding the chore of entrepreneurship which he either hates or lacks the courage to undertake.

The key to a private inventor's success is his ability to sell his invention, which involves technological innovation applied to some real need. As a private inventor, he is outside the industrial product chain described above and therefore has no access to the built-in "needs" information of the chain. Thus, to keep the information channel open, he probably has to build up a clientele of manufacturers. Once he has established a reputation as one who can deliver, people will bring their "need" requests to him and expect in return a right of "first refusal" when a new invention is generated.

A few examples of private inventors are presented here as reference:

- The "Gripnail" by R.L. Hallock*

  Knowledge of Needs

  "In 1955, when the author (Mr. Hallock) was returning from a convention in Chicago, his seat companion, Jules Hollman, Vice President of the Flintkote Corporation, remarked that his company has been unable to find a satisfactory way to secure fiberglass insulation to the inside of sheet metal air ducts. Hollman wondered

---

Inventing For Fun and Profit, Robert Lay Hallock, Harmony Books.

whether the author could do something about this problem, knowing that he had developed several fastners.

Air-conditioning ducts are used in large buildings and in private homes to carry hot and cool air to various rooms. To eliminate heat loss while the air is traveling the ducts, insulation is wrapped either around the outside of the air duct or secured to the inside. The insulation material is almost always a fiberglass type of material. It is ½ or 1 inch thick and in density is about like a mat of cotton with a thin film of paper on one side. It is so loosely bound that when you press down on it, you can compress it to a thickness of ¼ inch.

It is now customary to insulate most air ducts by placing the insulation on the inside of the air duct. This avoids tearing of the insulation in shipping from the place of manufacture to the point where the ducts are to be installed.''

Define the Needs

After some study, Mr. Hallock realized that the following earlier attempts had been unsatisfactory:

1. Cementing the fiberglass insulation directly to the metal surface; this is not satisfactory because the fibers are laid in layers so that cementing the inside layer would not prevent the outside layer from separating.
2. Securing the fiberglass by a nail with a large flange driven through the fiberglass and the sheet metal; the flange is satisfactory, but penetrating the sheet metal is undesirable, because it causes air leakage, and the protrusion of the nail might interfere with the surroundings.
3. Placing the fiberglass over a stud which is pre-cemented to the sheet metal with a metal clip then clamped over the stud; this is functionally acceptable but too costly.
4. Welding the fastener is too complicated and causes dispersion of the protective zinc coating.

With the above schemes having been rejected, he then defined his goals:

1. the fastener should be in the form of a flanged nail;
2. use a simple tool like a hammer to attach the nail to the sheet metal;
3. the nail should not penetrate the sheet metal but enter the metal partway and then be deformed at the tip to develop an expansion joint.

Prior to this analysis, Mr. Hallock had been successful with his sheetrock fastener invention. In that device, his fastener bent over behind the hole of the sheetrock to achieve the grip. His past experience and achievement prompted the conversation with Mr. Hollman and gave him a grasp of the problem and established his alertness to a perceived need. A review of all unacceptable solutions led him to draw on his knowledge of fastener design for sheetrock and inspiration was born, a process which may have taken only one or two hours or a few days when the problem was still fresh in his mind. It is interesting to note that after the need was perceived, it then took him five years to complete the development of his "gripnail," about which he commented:

> "The task of inventing the 'gripnail' required five years of experimentation and hundreds of trials. Any effort to gloss over the expenditure of time and the total frustrations would be a gross disservice to the would-be inventor. Likewise, any detailed retelling of the exact series of trials and false leads would negate the narrative value of this booklet."

The final form of his gripnail is shown in Figure 5.3. Interestingly, there are some similarities between this innovation and the wire wrapper described earlier: Both systems involve mating two parts by deformation. In wire wrapping, the bending is effected by tension, while here bending is effected by compression. Some detailed study reveals that:

1. The core of the hollowed portion of the nail serves as the stop to limit penetration.
2. A groove (see Figure 5.3) is provided at the root of the "lip" to minimize bending stress.
3. The "ring" shaped lip is divided into two halves. This provision

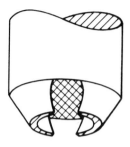

Fig. 5.3. The tip of the gripnail.

would break up the hoop tension in the ring which prevents the lip from curving in. On the other hand, if the lip were divided into too many sections, then each section would be too flimsy as a "column" to transmit compressive forces.

4. The "bevel" is placed on the outside of the lip to force the lip to bend inward. The grip thus generated allows the nail to tilt sideways over a significant angle from normal to the sheet without coming off. This is because the "bite" thus created would allow the lip (or tooth) to be pulled along the direction of indentation (Figure 5.4) without being straightened; in this manner, it can be subjected to alternating applications of stress without losing the bite.

Conversely, an expanding lip would be bent straight when the nail is tilted sideways slightly, as shown in Figure 5.5. This type of bite offers stronger resistance to the first tilt but loses the bite quickly after being rocked once or twice. For this kind of application, there is no need to maintain lateral rigidity of the nail, and freedom in tilt motion is indeed preferred, taking into consideration the vibrations the conduit might have to face.

7. THE CRITICAL MOMENT IN FOUNDING A COMPANY— IDENTIFYING A NEW MARKET NICHE

Every new industrial firm is organized on the founder's belief that he has a product that will satisfy a need in order to fit into a market niche. This is a very crucial perriod, because the new business has had no cus-

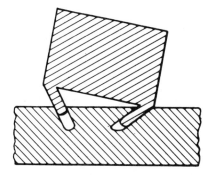

Fig. 5.4. Properly shaped gripnail.

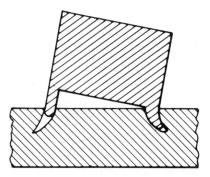

Fig. 5.5. Improperly shaped gripnail.

tomers and few alternatives. Aside from tightening the belt and hard work on the part of the founder and his associates (which everyone takes for granted), the ultimate proof rests with the soundness of his initial estimate of the need and the availability of such a market niche. If the company has sufficient resources, the founder may have the option of changing horses in midstream until the market niche for an alternative product is found.

In this section, two examples will be used to illustrate this crucial'' founding'' interlude of two companies; both get over the hurdle with a combination of luck and skill.

Case 1. Kybe Corporation

Kybe is a small industrial firm in metropolitan Boston. It was founded by Dr. Way D. Woo in the sixties, when he decided to leave a large company to become his own boss. His initial product was an ingeniously conceived curve reader, which allowed the user to manipulate any geometrically shaped curve and have it instantly converted into digital encoding. The need for a curve reader was probably recognized by him through his association with a large, industrial firm. It was a time when forming a new company was easy, and his belief in the new enterprise was sufficient for him to leave his good job and raise enough capital to found Kybe. However, the market for the curve reader did not materialize; by coincidence, a neighboring firm needed a device to clean magnetic tape, and Woo and his associate took up the challenge and built a simple device which drove the "dirty" tape over a carbide blade to scrape off the excessive ferropowder aggregated on the surface of the tape. (This unsecured deposit tends to give erroneous information and is annoying to the user.)

The mechanical scraping device appeared to work, but it needed an instrument to give the tape a "before and after" type of verification in order to prove the effectiveness of the scheme or any other cleaning process.

Woo exercised his innovative abilities and developed a tape certifier, which subjects the tape to a known, random signal and then uses the correlation between input and playback signals as an indication of its cleanliness. This tape cleaner and certifier turned out to be a blessing for Kybe, because they satisfied a need, while the original innovation—the curve reader—never took off. With the ever increasing quantity of magnetic tapes in use, it was not hard to convince the user of the reliability of recycling tape at a fraction of the cost of buying new tape. The manufacturing business flourished, and Kybe soon branched into the "tape laundry" business through franchise arrangements.

Case 2. The ECD Corporation

Richard Eckhart was determined to be an entrepreneur even when he was an M.I.T. sophomore. His participation in the first innovation center class offered during his Junior year hardened his will. Thus, as soon as he finished the B.S. program, he and two other friends made the plunge and organized the ECD Corporation. They selected a power supply as their initial product but quickly found the market for that device to be quite saturated.

In the course of engineering development work in their shop, they found a need for a capacitance meter, though it was debatable how many people would purchase such an item. It is not needed so often as voltage and resistance meters, but it was also true that very few capacitance meters were on the market at the time. As an example, Setra Systems, Inc., a capacitive transducer manufacturer, had developed capacitance meters for their own use, yet it had never occurred to them to market their capacitance meters, because it did not appear to be so challenging or necessary as the transducers; Setra was wrong in this estimate, however.

Disillusioned by the power supply, Eckhart and his partner thought that a capacitance meter might be a product for ECD, partly because they sensed the need in their own work, and partly because building a capacitance meter had been the subject of Dick Eckhart's bachelor's thesis. On the basis of their early experience, they set as their goal the design of a capacitance meter employing the latest technology and well planned, industrial engineering techniques to offer the user ultimate convenience.

Noteworthy features include:

- a large three-digit display;

- a self-ranging system to cover six ranges from 100 pf to 100 μ;
- a quick plug-in for different types of capacitors;
- a long life prime cell instead of a rechargeable battery to void the need for frequent recharge:

ECD even supplied the user with a detailed circuit diagram, thorough instructions, and a booklet highlighting many interesting features about many types of capacitors. This instrument was an instant success. In two years, the annual sales volume reached one million dollars. There were many unforeseen users, including utility companies, who gave the meters to their linemen to diagnose problems with transformers on utility poles.

In the above two examples, we see two earlier failures followed by two later successes: The digital curve reader invented by Woo was clearly a limited market device; the power supply attempted by ECD, on the other hand, entered a very congested market, where it is extremely difficult for a newcomer to establish his identity and thereby achieve a foothold. In such cases, it may be suggested that there was no need for a newcomer, instead of saying there was no need for that product.

The tape cleaner and certifier and the capacitance meter are excellent examples of products filling a market niche. The tape cleaner was a chance encounter for Kybe, but the follow through with the certifier was an innovative move. It also fully illustrates the essence of needs analysis: Unless the user's confidence is developed by observing the effects of the cleaner, how can he be persuaded to employ it. EDC's capacitance meter is an excellent example of thorough user oriented, engineering design which gives instant satisfaction. The designer should also be congratulated, because no design modification was needed after the introduction of the initial model; not a bad extension of a bachelors thesis!

3. AN INVENTOR'S DREAM TO BE FIRST WITH A SIMPLE AND POPULAR SCHEME

The random-access, magnetic core memory scheme, invented by Professor Jay Forrester of M.I.T. and yielding a total royalty payment in eight figures for M.I.T., is certainly an eye opener. The need was obvious at that period of time, and the solution was elegantly simple.

Another invention equally appealing is the dual slope analog-to-digital converter. According to Bernard Gordon:*

Linear Electronic Analog/Digital Conversion Architectures. by B.M. Gordon, an Analogic Corporation booklet.

"While the dual slope structure as shown at first glance uses the same building block elements as do the previously discussed charge replacement type converters, it represents a major, elegant breakthrough in concept. Unlike the previous types, it is easily auto-zeroed. But of greater importance, it provides an output code independent of both the clockrate and the time constant of the integrator network. A properly designed dual slope converter can therefore have excellent zero stability, excellent linearity (including differential linearity), and an absolute accuracy, independent of almost every component, except of course the reference accuracy. The dual slope converter has made a major imprint on low-speed precision instrumentation."

Need for a good Analog (digital) converter was clearly defined by Gordon. The elegant solution is indeed quite simple, it involves:

- an accurate $\pm$ reference voltage V_r with which the unknown voltage V_m is to be compared;
- a few accurate electronic switches;
- a clock counter which does not have to have absolute accuracy but should be even among the beats;
- an integrator which generates an output signal V_0 whose rate of change is proportional linearly to the input voltage

$$\dot{V}_0 = aV_{in}.$$

The sequence of operation involves:

- Integration of the measured voltage V_m until the output V_0 reaches a certain voltage, V_c; the time T_m of this integration process is established with the clock count.
- Use of a reverse voltage reference to discharge the integrator to bring the output voltage of the integrator from V_c back to zero; the discharge time T_R is also established with the clock count.
- Computation of the measured voltage V_m as:

$$V_r = \frac{T_m}{T_R}.$$

Those with a general electronics background would see quickly that this is a good scheme under the condition that the needed components, with the desirable accuracies, are available (this was possible at the time the scheme was invented). Just imagine, only about 15 years ago, the Air Force or Navy would pay 20,000 dollars to acquire an accurate A convertor, which now costs only a few dollars. The dual slope A/D convertor patents are now owned by the Western Electric Company. The list of Gilbert and Ammann Patent licensees covers 42 companies, including Analogic, Analog Device, Hewlett-Packard, Fairchild, I.B.M., and Triplett, to name a few.

9. THE MULTIDIMENSIONAL NATURE OF PRODUCT DEVELOPMENT

The growth pattern of new product development shown in Figure 5.1 is a simplification of the general situation, assuming that a new product originates from one dominant innovation. In complicated systems, such as the C.A.T. scanner discussed in Chapter 3 the multinational company that produced and marketed the system sought technological innovation from Analogic, a smaller company, to reinforce the producer's own long established experience in conventional x-ray machines. In the development of the word processing system, Wang Laboratories also supplemented its own innovation with that of others through acquisition; likewise, Analog Device strengthened its in-house, large scale, integrated circuit capability through the joint venture approach.

In all these cases, the need perceived by the respective management teams was not just an isolated case of market niches but to a large extent reflected the company's own need for full utilization of the existing marketing and production capability. This is why Ray Stata defines his innovation climate by giving each innovator in the company certain freedoms but within specified areas. In the case of Wang Laboratory, the expertise of the 2600-man sales force is likely to dictate the immediate area for expanding activities. Thus, for the present period, innovative marketing strategy is equally, if not more, important than technological innovation for the improvement of profit and the growth of the word processor.

The entry of Exxon into the word processor market and the emergence of Kodak in Polaroid's instant photography field represent still another level of managerial confidence that a selected market can be captured, starting from scratch with a massive backing of capital; indeed, this is possible and a familiar pattern after a fertile market has been opened up by a new breakthrough in technological innovation. It is for all these reasons that each new product must be developed fast enough to secure a significant market base before being squeezed out by a stronger intruder. However, it is equally prudent to go slowly in order to try out all the alternatives suggested by Wang in the last chapter. In this respect, the "uniqueness" of the innovation became important. A strong basic patent, like the radial tire, the dual slope, analog digital converting scheme, the Polaroid film, the magnetic memory core, the grip nail, Analogic's deblurring scheme for the C.A.T. scanner all seem to be simple and elegant with long lasting and protected markets.

All these innovations were derived from the perception of a need and developed through the search for solutions, illustrated in Chapter 3. It is believed that with some training in parameter analysis methodology skills can be developed which will lead to a higher yield of better quality innovations.

6
GENERATING NEW IDEAS

David G. Jansson

1. INTRODUCTION

Innovation is the introduction of something which is new or different. In this book, we are particularly interested in technological innovation and the introduction of a new product into the marketplace to satisfy the needs of society. Innovation is a much broader term than invention, the latter being used to describe the process of creating something new. With this understanding, it is clear that invention is but a part of the whole innovation process.

Figure 2.1 of Chapter 2 depicts the technological innovation process. Stations II and III of this structure represent the process of creating or devising a technological solution to satisfy the need which has been identified at station I. The flow of the innovation process, of course, depends uniquely upon the generation of an idea which is capable of being the focal point of the process downstream. As introduced in Chapter II, parameter analysis is a central theme of our innovation methodology. Although parameter analysis principles are applicable to all phases of the innovation process, it primarily impacts the function of conceiving and selecting a technological solution which will satisfy the need.

2. INVENTION METHODOLOGY

We describe the methodology of invention or generation of new ideas with the following five important steps indicated in the structure of Figure 6.1. The input to the process is a recognized market need; the

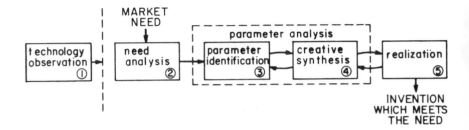

Fig. 6.1. The invention process.

output is, of course, an invention * which satisfies the specifications of the need. The new idea is then ready to move into stage III of the innovation process, as described in Chapter II. The particular function of each of these steps is discussed in this chapter, and examples are given to illustrate how this methodology is applied.

Before considering this concept in more detail, we must begin by recognizing that in the invention process, nothing can replace that spark of creativity which ignites every good idea. Some people are blessed with a measure of creativity and inventiveness which others will never be able to experience. We all have known or heard of individuals for whom invention is a natural activity, but this group is a small one and not representative of the vast majority of individuals involved in the innovation process.

Like any skill, invention takes practice to be developed to its full potential. The analogy of an athlete is particularly clear here. Good athletes are fortunate in having a high degree of natural ability; some, of course, have more than others. Nevertheless, any athlete, regardless of his natural gifts, must persevere through long, hard hours of practice to attain a competitively high level of performance. In a very similar way, to become a good inventor, one must "practice" continuously to develop his skills. Every coach has a method by which he trains his athletes. Thus the various steps of this invention methodology form a coaching method that can sharpen one's inventive talents and help the sparks of creativity to ignite with more regularity.

* NOTE: The word invention here merely indicates a new idea and is not necessarily restricted to the patent office definition.

3. TECHNOLOGY OBSERVATION

The observation of technology is the first step in our description of the invention process. Rather than being a specific step in the process each time a particular need is identified, it represents the base upon which the methodology rests. It is the continuous process of observing and analyzing the products of technology to familiarize oneself with what others are doing. The noteworthy feature of this observation is that the important elements of technology for the innovator are the how and why rather than merely what has been done. Thus, the base of knowledge which an innovator accumulates over the years consists in understanding the underlying features and concepts of a configuration or phenomenon rather than details of the design. In the observed application of technology to the creation of something new, it is an appreciation of these underlying principles, in contrast to design specifics, which permits us to draw proper analogies or gain insight into the problem and solutions at hand. The approach to technology observation is identifying key parameters and principles—analyzing what is critical to the successful operation of someone else's innovation. With the radial, belted tire, the principle of avoiding lateral rubbing on the road by radial, belted design is the dominant feature of the tire and a piece of information which may be more useful to the innovator than the mere fact that the radial, belted tire indeed lasts longer (see Chapter II).

The role of technology observation in the invention process will become clearer after a more complete discussion of the other steps involved. An example of applying technology observation is given below to help clarify the intended thrust of this preparatory step in the process of generating new ideas.

A simple device called a fluidyne is depicted in Figure 6.2. (The fluidyne heat engine was invented by Dr. Colin West at Hartwell in Britain in 1970.) The particular device being used for this example is made of ½-inch diameter glass tubing bent into shape with the tube partially filled with water. A short section of tube C is heated with an electrical heating coil. A few minutes after the coil is energized, the fluid begins to oscillate up and down in all three tubes; it quickly reaches stable oscillating conditions. This behavior is an excellent example of the importance of identifying the underlying principles while

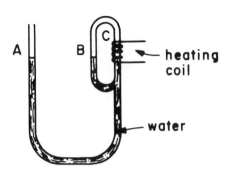

Fig. 6.2. Fluidyne.

observing a new device. It is interesting to note that the fluid indeed moves up and down and that mechanical energy can be removed from tube A, but, unless we understand why, this novel configuration would be of little use to us as innovators.

First, we have to observe the performance of the device. We note that the oscillations have a frequency of about one cycle per second and that the three columns have a phase relationship between them. Columns A and C are nearly 180 degrees out of phase, and column B oscillates with smaller amplitude, lagging behind C by about 90 degrees. While the heater is off, the water, when disturbed, will oscillate at about the same frequency as the operating frequency. After a very few cycles, however, the motion stops. During operation, one also observes condensation dripping down both B and C.

At first glance, we realize that the fluidyne is a very complicated dynamic device, involving thermodynamic processes, as well as mechanics. Energy is introduced into the system by an essentially steady source of heat, producing an oscillating, mechanical output from which energy can be taken. While avoiding a detailed, quantitative study of this heat engine, we would like to gain insight into what makes it go.

Since the phenomenon is clearly dynamic in nature, it is reasonable to concentrate on the dynamic characteristics of the fluidyne. Oscillating systems are common to all disciplines. We have already noted some important features, such as frequency, amplitude, and phase, but there are underlying properties of oscillatory motion which provide better insight into what is happening. Oscillatory motion can take place only when energy is being transferred from one mode of storage to a other. Thus, energy is stored by virtue of the motion of the fluid as kinetic energy; however, energy is being transferred from the moving fluid to potential energy of some form to permit the system to oscillate. Here there are at least two identifiable mechanisms: gravity and the elasticity of the gas in the closed volume above columns B and C, which store potential energy. To increase the amplitude of oscillation, energy must be added. Loss of energy from the system decreases the amplitude of motion. When the energy added to the system during one cycle is equal to that lost, the amplitude of the oscillations remains unchanged. Thus, if we can identify the mechanisms of energy gain and loss during each cycle, we will have identified some key factors in the operation of the fluidyne.

In order to add energy to the moving column, force must be applied in the same direction as the velocity. If the opposite situation is true, energy is lost from the motion. Even though energy in the form of heat is supplied by a steady source, since the motion of the fluid is oscillatory, a mechanism which provides force in phase with the velocity must be identified. Furthermore, since we know that the steady state frequency of the device is very nearly the same as the natural frequency of the undriven system, we know that the driving force we are looking for is applied for only a very short portion of the cycle. If this were not the case, the operating frequency would differ significantly from the natural frequency of the system. We expect to find a phenomenon which provides an inphase driving force for periods on the order of one-tenth to one-quarter of a second, since the frequency is about one cycle per second.

The only mechanism available for transforming into mechanical energy is a pressure rise and expansion of the gas in tubes B and C, but, as we have just stated, this phenomenon must take place rapidly. The surface of the water in column C moves up and down entirely within the length of the heater coil, so we do not expect boiling from this surface to account for the force. Furthermore, if the heater were merely heating the gas above the fluid and boiling off vapor from the fluid, this would cause only a static displacement in the column. We also recognize that since the thermal conductivity of glass is low and the thermal capacity of the heater and glass is quite high, the temperature of the tube is essentially constant during operation. The only phenomenon which takes place rapidly enough to account for the short pressure pulse necessary to drive the column of fluid is the wetting of the tube's surface by the liquid surface. The thin film left by this action allows evaporation to take place very rapidly, since, given the small volume of fluid, the ratio of thermal capacity to heat transfer area is very low. Now, when the wet surface is left behind, it evaporates almost immediately, raising the pressure of the gas in phase with the motion of the fluid, thereby adding mechanical energy to the system during each cycle and replacing the energy that is lost through friction or other mechanisms, such as column A doing work. The vapor condenses on the cooler portions of tubes B and C and returns to the fluid columns in a continuous fashion.

It might be useful here to recognize the utility of this configuration as a heat engine by pointing out that this single-stroke, two-cycle engine operates at a very low thermodynamic efficiency. Since the pressure-volume curve for a cycle differs only slightly from the line representing the pressure-volume characteristics of air, the work output represented by area enclosed within this curve is very small.

Although this is far from a complete description and analysis of fluidyne operation, we have uncovered the essential elements to point where we could use the configuration or draw upon our understanding of the phenomena in application to other situations. In the discussion centering around Figure 2.3 in Chapter II describing parameter analysis, the technology base consisting of innovation building block parameters is a central ingredient. In our example, we can identify several concepts which we should put in our bag of tricks. A complete physical understanding of the dynamics of oscillatory systems including the concepts of how energy is added, transferred, and lost an important piece of information to have at one's disposal. The unique ability of thin films to affect rapid evaporation is another essential piece of information. Although a portion of this may have already been known, some new understanding may result from observing the technology of others. We believe that the inventor should be continual learning in this fashion and depth what the surrounding technology teaches. In addition to bolstering his own technology base, such practice sharpens his skill in identifying critical issues in all phases of the innovation process.

4. NEED ANALYSIS

The second facet of the invention process is a careful analysis of the recognized need in order to determine more precisely what particular characteristics the new product is required to have. The following list of general categories forms a set of constraints which the new product must meet:

a. performance
b. cost
c. size
d. safety
e. market constraints

Although need analysis is sometimes a straightforward statement of the constraints listed here, oftentimes the analysis results in a restatement of the general need because of a new, clearer understanding of these constraints and their effect on a product in the marketplace. For example, trade-offs between cost and various elements of performance often clarify our perception of the real need, which may differ considerably from the statement as first perceived. When considering the development of a laborsaving device to replace either manual laborers or existing capital equipment with new technology, determining present costs to accomplish a given task will clearly define cost constraints on the new product. Redefining the task to be accomplished by the new device because of better appreciation for the present cost/performance balance may be a further result.

It is of paramount importance during need analysis to restrain yourself from inventing a solution before the need has been completely described. If this is not the case, your potential solution may put unfair or unrealistic constraints on the need specifications and either prevent the actual problem from being defined or a highly superior solution from being created. This is often a difficult task, but, with practice, you can develop skill in quickly specifying the essential constraints on a new product solution.

It should be pointed out that input to the need analysis process is a market need which has already been identified. The whole innovation process, including the specifics of invention that we are discussing here, depends on clarity and vision in perceiving the market. The results of a good need analysis can increase our confidence in a particular need definition, as much as it can discourage us from pursuing the problem any further. Nevertheless, at this very early stage in the innovation cycle, the risk is great, making it essential that our new idea be superior.

As a specific example of invention to follow through the various steps in sequence, we will describe a problem for which no solution exists in the marketplace, thereby avoiding 20-20 hindsight in our thinking. Let's assume that we have identified the need for a robot lawnmower for very large lawns. Although this represents a relatively complex invention problem, it will illustrate the methodology very well. We wish to invent a device or several devices which will automatically, with a minimum of human intervention, perform the task of normal upkeep for a large lawn, such as a golf course. As we have described, the proper way to approach this problem is to first perform a need analysis; the outline of this step for our example follows.

For this example, we will focus particularly on the problem of mowing both the fairways and greens of a typical golf course, since this specific set of customers probably represents a significant part of the potential market. In addition, the problems associated with golf course mowing comprise a reasonably stringent set of constraints which, if satisfied, will result in a more widely applicable product. Having identified more specifically a need we wish to meet with a product, we then determine what general specifications will constrain our solution. Some important factors which should be considered in determining these constraints are:

Performance, Size, and Safety

1. total area of fairways on an average golf course
2. total area of greens
3. time allowable for performance of the task
4. typical height of grass before and after cutting on both greens and fairways
5. special cutting requirements for greens
6. presence of irregular shapes, boundaries, contours and hazards
7. cuttings are/are not to be removed,
8. amount of cuttings produced
9. constraints of pathway size and strength; for example, device may have to use a bridge to cross over a stream
10. allowable weight of devices bearing on greens and fairways
11. maximum speed and stopping time of an unmanned vehicle for safe operation
12. general problem of safety with regard to the presence of humans or animals on the lawn while device is operating

Cost Considerations

1. number of people new equipment will actually replace, keeping in mind the necessity of maintenance and other human intervention
2. cost of present equipment to be replaced
3. value of superior performance if achieved
4. rate of return on inventment, if this is an option to a customer
5. cost of loan, if necessary, for customer

Using this probably incomplete list of performance, size, and safety considerations, with little or no additional information, even someone unfamiliar with golf and golf courses can make quite reasonable estimates of cutting specifications, allowable size of the equipment, speed required as a function of various other machine characteristics, and general, safety constraints on the new device. Although the specific numbers are not vital to our illustration, the following is a summary of one such effort at estimating performance, size, and safety constraints on our sample problem.

The fairways of an 18-hole golf course have an estimated area of 400,000 square yards and, if mowed twice a week during the night with a device designed to pass over a narrow bridge, would require a speed of only about 4 miles per hour; this is a reasonably safe speed of operation. The cutting could produce a few hundred thousand pounds of clippings which may have to be removed; the task of mowing the greens is small compared to the important constraints of cutting pattern and strategy.

The cost factors considered here basically revolve around the question of what we as inventors can expect the market to bear. This estimate is not meant to be the driving force of the invention process and should not become the overriding constraint; however, it is very useful before one spends a lot of effort on a particular problem to have a feel for the economic constraints on the solution. With very little effort, one can construct a graph, as shown in Figure 6.3, for our example which describes the relationship between cost to the customer and performance provided by the new device. For our example, we assume that the country club borrows the necessary money to purchase the robot system and pays back the loan with the amount saved by replacing a certain number of employees. There are many factors affecting such a curve which may not be easily estimated, and it is best

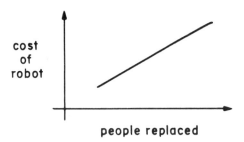

Figure 6.3. Cost/performance trade-off.

to underestimate what people will be able to spend if there is a great amount of uncertainty.

Since the club is borrowing money for the purchase, it must decide how long it should take for cost savings to balance the purchase price. This period of time is called the payback period, and there is a lot of pressure to make this very short for capital purchases, such as our example. If the customer chooses a payback period of two years and saves roughly 30,000 dollars per year in labor costs, he should be willing to borrow about 54,000 dollars to buy the robot. Implicit in this type of estimate is, of course, the effect of such factors as maintenance costs and expected lifespan of the equipment, but remember, we are making only rough estimates and have not even invented anything yet.

Including market constraints in our need analysis primarily reflects the fact that the need we are analyzing has been identified by perceiving the market, and some constraints are implicit in this recognition. As examples of constraints worth noting at this early stage in the invention process, we list the following:

Market Considerations

1. total market available for robots
2. competition
3. psychological and social factors affecting purchase, such as desirability of replacing present help, even if economic tradeoffs are favorable

Again, as in the case of cost estimation, unless such analysis yields insurmountable obstacles to success, it can often stifle the creative process to dwell on this area early in the game.

Recapping the need analysis of our example, we arrived at a list of

characteristics and constraints for the automatic lawnmower by using
estimates and reasonable guesses, while specifically avoiding trying to
invent a configuration which would do the job. (We have admittedly
excluded solutions, such as inventing grass which does not grow!)
Thus, we have prepared ourselves for the next phase of the invention
process. By skillfully analyzing the limiting factors which bound the
solution, we have a good start on generating an idea for a superior
product which can be successful in the marketplace.

5. PARAMETER ANALYSIS

The next two parts of the invention process join together to form the
crux of the methodology we are describing. Parameter analysis, as discussed in Chapter II, is a matching process divided into two,
definable steps, parameter identification and creative synthesis. Since
it was well developed in the earlier chapter, we will only review and
apply it, here in our example invention problem.

Parameter Identification

As the name implies, parameter identification is the process of recognizing the critical, overriding factors or variables which influence
the solution to a problem; almost equally vital is the ability to set aside
those parameters and effects of secondary importance. The parameters
identified during this phase are not necessarily familiar quantities or
phenomena but may represent a combination of effects which truly describe the central issues of the problem. It is because of this that parameter analysis is an iterative matching process, since attempts to create solutions based on identified parameters very often lead to refined
statements of key issues. The fruit of one's technology observation efforts, along with other experience which builds up a base of knowledge, enhance the ability to quickly discern the principle features governing a particular problem.

Considering our robot lawnmower, we recognize several features of
prime importance and many others which are not:

Primary Issues

1. navigation: The robot has to know where it is supposed to be;

2. accuracy: The navigation and guidance function above must be done with accuracy, as defined by specifications;

3. problem of handling grass clippings, if removal is desired.

Secondary Issues—We already know how to:

1. cut grass;
2. propel, brake, and steer a vehicle;
3. design vehicles for various terrain conditions;
4. solve many other design problems not central to our problem.

We will return to parameter identification after considering the rest o the parameter analysis technique.

Creative Synthesis

Even with the most careful and thorough preparation, the importanc of creativity in invention cannot be overlooked. Creative problem solving has received considerable attention in recent years, and, i comparing our invention methodology with other approaches, thes techniques seem to correspond largely to the creative synthesis phas of the process. Methods such as brainstorming and synectics * are e forts to increase the creativity of individuals through group interactio and experience has shown that indeed creativity can be enhanced b such methods. However, it is creativity in conjunction with a deep u derstanding of the problem which forms the critical combination nece: sary for generating truly unique ideas. Our approach to invention e hances this creative step in at least two important ways: Firstly, tl real problems have been lifted from the detailed goals of the task; se ondly, observing technology provides us with a level of understandi more easily matched to the important features of our problem.

A good way to begin the iterative process we are describing is identify the primary issues, as we have done in our example, and th make a list of as many general schemes or techniques that might pr vide a solution. By considering the list more closely, we can oft sharpen our statement of critical issues, which then helps us foc more clearly on relevant technology and configurations.

In our example, we will consider only the first two primary issu

* Gordon, W. J. J., *Synectics*, New York: Harper & Brothers, 1961.

relating to robot navigation and guidance; a list of possible schemes could include the following:

1. radar, microwave, or optical
2. preprogrammed onboard mapping
3. buried devices, active or passive

This is only a short list, but a more careful look highlights one particular parameter not yet identified: One property of the various schemes is that they vary in the amount of open-loop and closed-loop control being used. If the system knows only at intervals where it is or where it must steer, between these intervals it is controlled in an open-loop fashion and is very susceptible to disturbances. It seems very likely that on a fairway with severe contours, wet and dry conditions, hard and soft soil, and other properties which can cause changes in the path of the vehicle, the robot will have to know at all times how far it is from its desired course. Therefore, we can refine our statement of the navigation problem to include closed-loop control of the robot. The characteristics of open and closed-loop systems should be information that an innovator has as part of his bag of tricks; next time the inventor will identify it even sooner!

In a similar fashion, the iterative matching process continues in search of the creative solution which best satisfies the specifications. The mix of identified parameters, the inventor's bag of tricks, and existing technology produces a fertile environment for the birth and growth of a creative solution, and the key to this process is clarity of insight in identifying the underlying parameters of the problem.

Although there are very likely several good ways of providing the navigation and guidance function for our robot, one specific solution which seems to meet the specifications very well is briefly described. In order to provide a closed-loop control signal all of the time, the path which the robot must follow is completely marked by burying a metallic line a few inches below the surface. The robot is then provided with metal detectors which can sense the presence of the line and send a control signal to the robot. A simple "planting" device can be designed to lay the line behind a tractor; very inexpensive line might even be made by burying old magnetic tape.

A solution has been found which satisfies the most important features of this need. Further work on the robot vehicle, of course, is required to complete the story, but the secondary requirements can be met principally with good engineering design practice. It should be pointed out here that the idea we have generated for our new product probably cannot be considered as an invention but rather as the application of prior invention art to a new product. This does not reduce the effectiveness or importance of the methodology but merely serves to illustrate the scope of the methodology in generating new ideas. The more detailed example at the end of the chapter illustrates an application of the methodology which clearly resulted in an invention.

6. REALIZATION

The last important phase in our structure for invention is the step called realization. The emphasis here is not on the design and construction of the entire invention or system but rather on experimentation which will determine whether or not the concept is feasible. If it is found to be feasible, it is ready for the design and development stage which follows; if not, we must return to the parameter analysis stage once again, as the arrows in Figure 6.1 indicate.

The experiments should be done as simply and inexpensively as possible and should be designed to test only those features of the invention which are critical to the demonstration of viability. Thus, in our example, we should consider testing the detectability of magnetic tape through a few inches of soil, since the length of line required for a typical golf course is long enough to render the cost of line significant if an inexpensive substitute, such as used magnetic tape did not work. We might also need to determine the corrosion properties of the line and correct any problems in this area.

We have outlined a methodology of invention effective in improving one's inventive capabilities; an example has been outlined to illustrate the application of the methodology to an identified market need. The remaining part of this chapter consists of a more detailed example to further illustrate the various steps involved in the invention process.

Application of the Methodology for the Invention of a Firefighter's Respirator

The Need

In the early seventies, there was heightened interest in the dangers involved in firefighting and the need for better safety equipment and strategies. The need for improved protective respiratory equipment was particularly evident. Firemen are exposed to stresses of many kinds, including extreme temperatures (both heat and cold), physical and emotional stresses brought about from being in a dangerous situation, and the hazards of particulate, gaseous, and vapor combustion products in the breathing environment. All of these factors have an effect on the breathing pattern and needs of firemen.

Need Analysis

Because of the extreme conditions mentioned, many of the requirements relating to performance, size, and safety could easily be determined by learning the experiences and various problems of firemen with present respirator equipment. Typically, a fireman needs one hour of breathing protection each time he uses a respirator. Working with many pounds of bulky gear, even without a respirator, movement is slow and difficult. A firefighter generally allows about 15 minutes to reach his desired location. Thirty minutes of activity in a burning building is considered as much as any individual can tolerate; at that point, a timer alarm warns the fireman to begin leaving the fire; again, this may take up to 15 minutes.

Total weight and weight distribution are important from the standpoint of extra effort required and changes in the user's balance; in addition, firemen must very often pass through small openings in buildings, so that size and contour of extra equipment carried should not prevent them from performing their tasks. Visibility through the facepiece, temperature of the inhaled gas, and the degree of interference to the user's breathing pattern are also factors which firemen feel are important.

There are many types of protective respiratory gear available to the firefighter. Some of these systems supply their own oxygen, while others provide only filtering. Due to the fact that a fire may result in a depleted oxygen supply and there exists a wide variety of combustible products, it is essential for respirators used by firemen to contain their own oxygen source. Consequently, the two most common devices used are compressed air and oxygen rebreather systems; other types have almost no firefighting application.

Compressed air systems consist of a large tank of compressed air which is strapped on the user's back and pressure valves and regulators controlling the flow of air to a facepiece. The facepiece contains a passive exhalation valve through which air is expelled after each breath. Oxygen rebreather systems consist of a compressed supply of pure oxygen, a highly compliant breathing bag, and a chemical scrubber for removing carbon dioxide from the exhaled gas. By means of tubing and passive valves connected to a facepiece, the user inhales gas directly from the breathing bag and exhales back into the bag through the scrubber. The compressed oxygen is usually bled very slowly into the breathing bag to replace oxygen consumed by the body. Figure 6.4 presents functional schematics of each of these systems as a useful reference.

Figure 6.5 presents typical specifications of these two respirator systems and indicates the major deficiencies associated with each. The size of the air tank in compressed air systems often makes movement awkward for the user, but the principal problems are weight and the limited air supply. A large fireman whose work output is high for any length of time is able to breathe the tank down to the point where the exit alarm is triggered in about 14-18 minutes. This is a serious limitation to the fireman and to the overall strategy of fire-fighting.

Oxygen rebreathers present a different set of problems: The gas contained in the facepiece is nearly 100% oxygen, and the risks associated with this situation are high. (Inevitable leakage and careless use rapidly deplete any initial supply of nitrogen.) For example, firefighters are often caught in situations where visibility is essentially zero, and the surrounding objects are burning; an unseen object, such as a post or falling beam, could easily dislodge the facepiece and expose them to the threat of rapid, high temperature combustion. (Breathing oxygen at elevated partial pressures does not present a problem for periods as short as those involved in firefighting; for prolonged periods such as use in mines, physiological constraints become an important factor. The temperature of the inhaled gas in an oxygen rebreather is often high due to the heat produced in the scrubber by absorption of both carbon dioxide and water. This elevated temperature puts an additional load on the respiratory system of the user.

Parameter Identification

One result of need analysis is that many of the specifications describing a problem and its solutions are given in terms of important parameters. However, the purpose of the parameter identification step

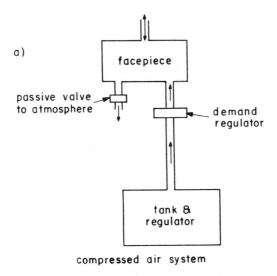

compressed air system

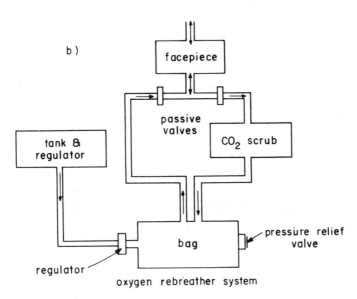

oxygen rebreather system

Figure 6.4. Functional schematics of two types of respirator.

respirator	weight	approximate size and shape	typical operating time (minimum)	major problems
standard compressed air systems	33 lbs.	22" 7" dia. cylinder	14-18 minutes	weight operating time
typical oxygen rebreathers	20 lbs	15" x 18" x 6" backpack	1 hour	safety weight temperature

Figure 6.5. Table of respirator specifications.

is to further identify those relationships which more effectively describe the salient issues. In this example, we have already recognized that weight is a very important constraint, we have also recognized that available operating time is critical for effective respirator utilization. We can see that there is a direct trade-off between these two factors in the present systems. In addition, we note that the oxygen rebreather system reaches a far more acceptable weight-operating time trade-off point but introduces new constraints, relating principally to operating safety. These parameters adequately describe the problem but from the innovator's point of view shed little new light on possible, innovative solutions.

By examining in a bit more detail how these existing systems work we can identify a set of parameters which better describe the respirator problem. Typically, about 10% of the oxygen inhaled in a breath of air is absorbed into the bloodstream, and a corresponding amount of carbon dioxide is transfered into the lungs. If the gas inhaled is not air but contains an elevated concentration of oxygen, the amount of oxygen absorbed is still about the same. Therefore, in compressed air systems, air containing 20% oxygen is inhaled, and air containing about 18% oxygen is exhaled into the environment. In oxygen rebreathers, as long as the pressure relief valve on the breathing bag remains closed, no gas is lost to the atmosphere. These observations combined with the necessity of keeping oxygen concentration in the facepiece low, provide us with a set of parameters characterizing the problem in a clear, effective way.

1. A key parameter affecting the weight-operating time trade-off is the ratio of the amount of oxygen utilized by the body to the amount carried in the tank.

2. A key parameter affecting operating safety and weight-operating time trade-off is the efficient use of the nitrogen supplied to control oxygen concentration in the facepiece. These two factors, efficiency of oxygen and nitrogen utilization, describe the problem in a manner which suggests new directions for invention; that is, we wish to provide a system which carries as much oxygen as possible with as little nitrogen as possible and allows a maximum amount of this oxygen to be used, while controlling oxygen concentration by means of the available nitrogen.

Creative Synthesis

The newly identified parameters immediately suggested a solution to this inventor. Since gas supplied for breathing should not contain less than about 20% oxygen and since typical usage demands that we be able to replenish our supply of nitrogen, any new system must have the ability to discard gas containing this lower limit of oxygen and retain gas still yielding available oxygen. A new system which provides the desirable characteristics is presented below; Figure 6.6 is a functional schematic of the new respirator.

The new system is a modified rebreather, which uses an oxygen-rich source in place of a pure oxygen source. System operation centers around the idea of rebreathing oxygen-rich gas from a breathing bag until the concentration of oxygen, $c(O_2)$, is low enough to be discarded. This "old" air is then vented to the atmosphere, and a fresh volume of gas is supplied. This operation causes the concentration of oxygen to have a time history, as shown in Figure 6.7. Each breath lowers $c(O_2)$, until the lower limit is reached. In order to avoid significantly long periods of time when the user might inhale without receiving air, two identical breathing bags are used. While one is being filled, the other is vented. These simultaneous filling and venting operations are conveniently done by forcing the volume of the bags into

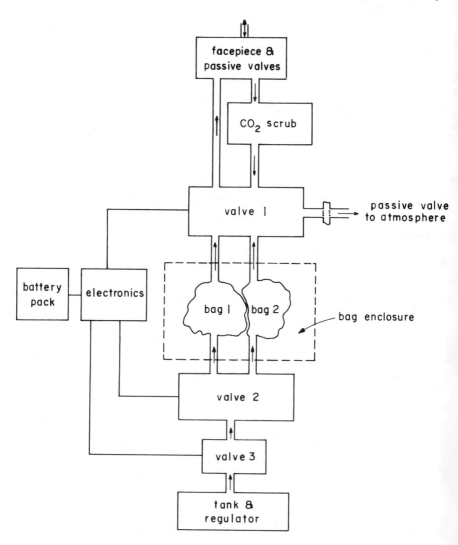

Figure 6.6. Schematics of a new system with oxygen enriched air.

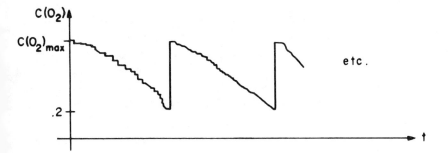

Figure 6.7. Time history of $C(O_2)$ in facepiece.

a single, stiff enclosure having the volume of one bag. The expansion of the bag being filled pushes the old air through a passive valve into the atmosphere, and energy to expel this gas is provided by the compressed supply.

The three numbered valves shown in the schematic diagram facilitate the above operations: Valve 1 chooses which bag is to be used for breathing and allows the other to be vented; valve 2 chooses which bag is to be filled; and valve 3 fills the bags with a preset volume of supply gas. There are a few variables which affect the operation of the system, such as the oxygen concentration of the supply, (R_s), and the fraction of bag volume added during each fill, (R_B). An analytic description of the operation of the system results in the following relationship between the maximum value of $c(O_2)$ and the system design parameters, where it has been assumed that switching from one bag to the other takes place at $c(O_2) = .2$.

$$c(O_2)_{max} = \frac{.2\,(1 - R_B) + 2R_sR_B}{R_B + 1}$$

By not completely filling each bag, we can choose the level of $c(O_2)_{max}$ necessary for safety reasons. The maximum oxygen efficiency α_{max} is defined as the ratio of the amount of oxygen available to the user to the total amount of oxygen in the supply tank and is determined by:

$$\alpha_{max} = 1.25\,R_s - .25.$$

again based on switching bags, when $c(O_2) = .2$.

Determining when to switch breathing bags is critical: The accuracy used to measure or estimate the point where $c(0_2) = .2$ will, of course, affect the efficiency of the system and the value of $c(0_2)_{max}$ obtained. In addition to measuring $c(0_2)$ with an oxygen sensor, less expensive and more reliable schemes include counting breaths, integrating total inhaled flow, and simple timing. All other methods approximate the function of the oxygen sensor and require adjusting the system to tailor it to the specific user. For short lengths of time, such as in firefighting application, these approximate methods are quite adequate.

Realization

The purpose of the realization step is to experiment sufficiently to determine feasibility of the new concept; in this example, there were two questions which must be answered:

1. Does the switching operation interfere with the user's breathing pattern? Any interference is unacceptable.
2. Could the valving and control functions be designed without extreme complexity and to operate reliably?

A laboratory prototype system was constructed with nearly all of the effort going into designing and building the valve and control functions. The experimental device demonstrated that the user's breathing pattern was uninterrupted and that, with relatively little complexity, the control sequencing could be accomplished.

A new scrubber was designed for the prototype which provided greatly improved temperature performance for a rebreather system, but discussion of this additional feature is outside the purpose of our example. In reviewing the invention process for the new respirator, the key to the creative solution was recognizing the unique way in which weight, operating time, and safety factors were related to the efficiencies of oxygen and nitrogen utilization. This new insight provided the spark for the creative step.

7
THE INNOVATION
PROCESS FROM THE
JUNCTION OF WOBBLING
DRIVE AND "HEAT MASS
AND MOMENTUM
TRANSFER"

1. INTRODUCTION

The concept of building block parameters was introduced in Chapter 2 (parameter analysis). In a broad sense, building block parameters encompass many technological as well as configuration parameters; to a lesser extent, they may include natural science parameters and consolidate specification parameters (Figure 2.3 of Chapter 2.). One way to acquire knowledge of building block parameters is through exposure to interesting phenomena so that the parameters may be identified and, thus, one's "bag of tricks" enlarged. But identifying interesting parameters is only half the innovation process: The complete process involves synthesis, or matching, the need parameters to certain building block parameters; this repetitive iteration of parameter identification and synthesis is parameter analysis. While skill in parameter analysis is the objective of innovation training, difficulty arises in the interface between the processes of teaching and innovation: The most efficient way of teaching is to drill well organized material into the learner's mind, step by step, whereas innovation is self-motivated, with emphasis on the thinking process itself and not necessarily on the material being used. Therefore, while a rigid structure may be an efficient means of transferring knowledge, over rigidity tends to inhibit the thinking process. On the other hand, without structure, it becomes difficult to put the knowledge or process into proper perspective and thereby establish a communication link for teaching and organizing one's own thoughts.

In this chapter, such a perspective is attained by identifying the "junction point" of well organized technology—"heat, mass, and momentum transfer"—and an interesting system configuration—" wobbling drive." From this junction point, various connections will be made to various, possible needs. In a broad sense, while this study is being conducted, searches will be made for certain higher level parameters beyond the new application just identified.

Figure 7.1 shows the apparently rigid structure of conventional engineering education, with its well organized hierarchy; on the other hand, Figure 7.2 shows the less regimented "junction point" structure. In this diagram, point *A* represents a junction between a well organized technology—such as the heat, mass, and momentum transfer phenomena (line I)—and an interesting system configuration, wobbling drive (line 1). From point *A*, several minor building block parameters radiate, and from these secondary parameters further connections will be

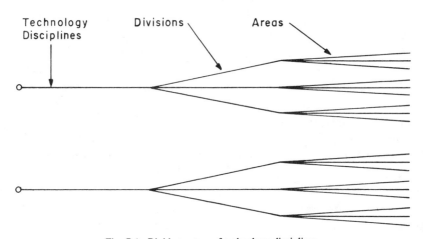

Fig. 7.1. Rigid structure of technology disciplines.

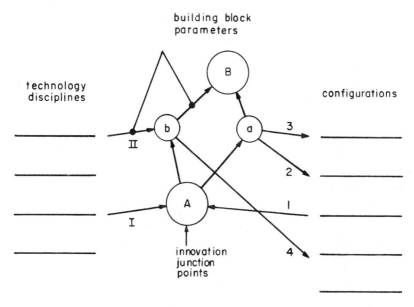

Fig. 7.2. Building block parameters.

made with other system configurations, such as lines 2, 3, and 4. Parameters *a* and *b* may have certain common characteristics, represented by paramater *B*. In this situation, parameter *B* would be one level higher in its hierarchy than parameters *a* and *b* and command broader applications than either of the parameters.

Junction point structure is a regenerative process which permits both synthesis and deduction. It may grow into a network with greater freedom than the rigid structure in Figure 7.2, and yet it still retains a definite form. Since this network is anchored to both the existing technology on the one hand and the existing configuration on the other, it can reach either end to obtain information resources or to innovate.

2. REACHING OUT FROM HEAT, MASS, AND MOMENTUM TRANSFER TECHNOLOGY

The philosophy of reaching out from an established branch of technology to a junction point in the building block parameter zone was described in a memo from Y.T. Li to his collaborators Cravalho and Jansson. An excerpt of that memo is herein reproduced:

As an attempt to examine this point of departure or point of convergence, let me describe my thoughts after having glanced through an excellent textbook, *Heat, Mass and Momentum Transfer* by Rohsenow and Choi. I found this book on my son's desk when I was visiting him yesterday at his fraternity house. In Chapter VI, "Heat Transfer in Stationary Systems," I was fascinated by a few interesting topics, such as section 6.7, "Critical Radius of Insulation," "The Center Line Temperature" in section 6.8, and "The Thermal Gradient Diagram" in section 6.9. These are all very interesting technological phenomena serving as excellent building blocks for innovative configurations. Thus, in a way, they may be considered the focal point where the science based technology treatment, was laid down in the book, and the innovation oriented parameter analysis, as we are trying to develop, intersect.

As do most excellent science based, technology textbooks, Rohsenow and Choi began with the elemental blocks, the equilibrium conditions, and then proceeded to lay down the differential equations and the boundary conditions. From these general and unified conditions, the elegant solutions evolve. These elegant equations may appear to be frightening for uninitiated persons, but they are really quite powerful, for when they are applied to certain restrictive conditions, the simplified equations would show the characteristics of the various applications very clearly. Again, as in all good textbooks, Rohsenow and Choi provided students with sets of useful homework problems on which to practice.

For instance, in section 6.7 of Rohsenow's book, "The Critical Radius of Insulation," it states that the heat loss of an insulated pipe may increase at first with the thickness of the insulation until the thickness reaches that which corresponds to the critical radius. Thereafter, the heat loss will decline with the increasing of the insulation, as one would expect. This example is interesting, because it contains the element of surprise. The objective of the book is, however, to demonstrate the "universality" of the treatment. This is then reinforced in the student's mind after observing the fine examples of practical applications and their performing of the exercises. The element of surprise or any of the other interesting features of the background of an illustration is incidental to the ultimate goal of proving the universality of the general equations and basic principles.

It is interesting to note that modern physics is built on elementary particles, chemistry on atoms, biology on molecules and radicals and engineering science on elemental blocks. For some engineering science, such as the "Mass Flow and Energy Transfer Theory," particles, atoms, molecules are too small to be concerned. The aggregation of elemental blocks, on the other hand, can indeed satisfy most of the physical embodiment where this particular engineering science applies.

The function of technological innovation is to generate a new and novel configuration to satisfy a social need. For instance, a certain configuration may include an insulated tube with certain physical parameters in conjunction with other physical components to perform specific tasks better than existing devices. Obviously, in designing the insulated tube, the information contained in Rohsenow's book is very helpful, but how to conceive a particular configuration prior to the call for the design computation requires some different kind of skill. For this function, the mental framework within which one's thinking process is organized would be different from the development of engineering science from the elemental blocks, the boundary conditions, and the differential equations; indeed, all these are parameters and parameter relationships relating to the final configuration. But, to play the game of innovation, we need a higher level of parameters which are closer to the final configuration than the elemental blocks, the molecules, the atoms, and the particles.

For instance, the reciprocal law of conductivity versus resistance (in one of the equations given in the Heat Transfer book) and all their implications are pertinent to the configuration. These laws also have commonality among several disciplines (e.g., electricity). Then there is the phenomenon of the ''interface heat transfer'' versus the ''internal heat transfer.'' These are interesting parameters associated with ''critical radius insulation.'' Working further with these parameters, one can see, for instance, that there is probably another parameter, such as critical tube radius. Beyond this radius, the critical insulation radius cannot exist for a certain ratio of the ''interface heat transfer coefficient'' to ''internal heat transfer coefficient.''

Engineering science was in full swing after World War II to refine the handbook approach where the physical properties of engineering building blocks were established by empirical rules. We are *not* trying to go back to the handbook era but trying to take off from both the handbook and the engineering science approach into a new dimension, where the pertinent parameters of the engineering building blocks are consolidated, broadened, and coordinated in such a way that the thinking process to generate new innovative configurations from the building blocks are rationalized and shortened.

3. REVIEW OF THE CONCEPT OF WOBBLING DRIVE

Wobbling drive is a rather unique mechanism which is, in effect, an extension of a four bar linkage, shown in Figure 7.3a and Figure 7.3b. The function of wobbling drive is to make the rigid frame A of Figure 6.3b wobble about a small circle with radius r while it remains parallel to the base. In principle, every point on the plane defined by frame A has its own center of rotation.

Figure 7.4 shows a very interesting device consisting of two spiral passages driven by wobbling motion with respect to each other. Depending upon the direction of the wobbling motion, the device may be used as a continuous expansion or compression device (patent managed by Arthus D. Little Inc. under the name of Scroll pump).

In the following study, wobbling drive as a means of moving liquid around by momentum transfer is of particular interest. For example, if one holds half a cup of tea and applies a wobbling motion to the cup, he will observe the tea spinning around inside the cup, while the cup

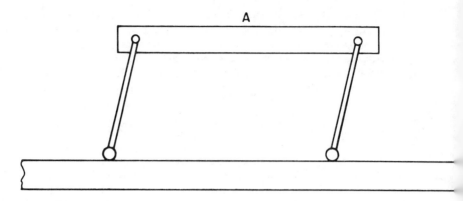

Fig. 7.3a. Four bar linkage.

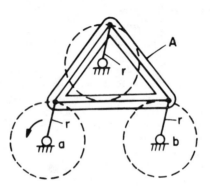

Fig. 7.3b. Wobbling drive.

Fig. 7.4. Concept underlying a "scroll" pump.

itself wobbles, not turns. How does the cup drive the tea since there is no revolving blade to push it? Would this simple and yet rather unique phenomenon serve any useful function? Some interesting applications of this kind of drive are herein described.

4. "ROTARY DRIVE OF A GASEOUS NUCLEAR ROCKET"

As an example of innovative ideas, Professor Kerrebrock described a gaseous, nuclear rocket (which he proposed some 15 years ago) to the Socio-Technological Innovations seminar. His scheme involved using nuclear fuel in gas form to heat hydrogen as the propellant. A conventional nuclear rocket has the nuclear fuel imbedded in a graphite structure, and temperature of the construction is limited by the property of the graphite. This limitation could be removed, if the nuclear fuel was utilized in gas form and contained in the combustion chamber by revolving at high speed, as shown in Figure 7.5. Assuming that the diameter of the nozzle of the rocket engine were considerably smaller than the diameter of the combustion chamber, it was hoped that nuclear fuel would be trapped within the combustion chamber by the centrifugal force of the revolving fuel. Hydrogen was also allowed to seep through the wall to form a boundary layer separating the nuclear fuel from the inside wall of the combustion chamber, therefore preventing the wall from overheating.

Originally, it was also hoped that nuclear fuel could be forced to whirl around inside the combustion chamber by the hydrogen introduced into the combustion chamber through small, tangential inlet nozzles. After a considerable amount of R&D work (costing millions of dollars), it has been found that there is not enough torque generated by the hydrogen to offset the loss of turbulence to the working medium revolving at the desired speed of close to Mach 1. Part of the difficulty lies in the small amount of hydrogen injected for propulsion purposes, partly due to inefficient momentum coupling between the hydrogen which was introduced and the nuclear fuel trapped in the combustion chamber. Finally, the dual purpose of utilizing hydrogen for tangential drive and boundary cooling seem incompatible. This particular example serves very well to illustrate the processes of innovation and acts as a focus for brainstorming in a seminar class. For this reason, the group devoted the next class hour to looking at the problem from different angles. One student suggested having the working flu-

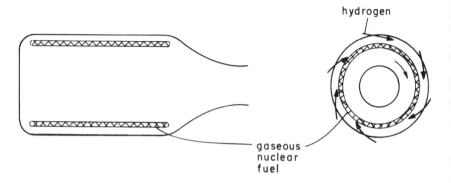

Fig. 7.5. Concept underlying a gaseous nuclear rocket.

id ionized, polarized, and then rotated by electricity. Y.T. Li thought it might be possible to introduce wobbling motion at the container wall in the same manner as one would whirl water around inside a teacup by shaking his hand rather than actually revolving the cup, an idea which was formed through the following two steps: The first, by recalling the concept of the explosive forming of sheet metal, which led to using a sequential pressure wave around the periphery of the combustion chamber to generate rotary motion. Then, on the evening following Kerrebrock's lecture, Li gave a talk on inertial navigation systems at a local IEEE meeting where he commented on an electrical inertial stabilization platform developed by a West Coast firm. This stabilized platform is in the form of a small, solid beryllium ball driven to rotation at very high speed inside a spherical cavity. The cavity was made of eight equal sectors of a hollowed sphere, and these sectors were electrically insulated and provided electrostatic coupling with the metal ball. The electrostatic coupling was used to generate suspension force, as well as to measure the gap between the ball and the electrostatic plates. In this manner, the ball could be suspended in the center of the cavity through a feedback arrangement and used as a seismic mass to measure acceleration in three dimensions.

Even more amazing was the use of the ball as a two-degree-of-freedom gyroscope by spinning it at very high speeds. To start the spin, electrical induction coupling was used; thereafter, rotation was sustained electrostatically. This was accomplished by having the geometrical center of the ball's surface offset from the mass center of the ball by a few microns in distance, thereby producing the inertia coupling of wobbling drive. Since the electrostatic force is always normal to the plate, as fluid pressure is normal to the container wall, the similarity between these two cases reminded Li of using wobbling drive for the gaseous nuclear rocket.

Indeed, the technology of rotary drive for the gaseous, nuclear propulsion system, as well as the parameters associated with the propulsion requirement for deep space travel fall into the general engineering science area of "heat, mass, and momentum transfer." On the other hand, several innovative, "inward driving" configurations are needed to contain high temperature media for various applications: Wobbling drive with a rigid container is one; the "magnetic bottle" concept for the nuclear fission process is another. This is just to show that there are several possible branches emanating from this junction point, but, for the sake of example, we limit ourselves to studying the feasibility of wobbling drive, assuming that by means of wobbling drive the temperature problem could be resolved if it were possible to achieve laminar flow for the layers of hydrogen and the gaseous nuclear fuel.

5. ANALYSIS OF WOBBLING DRIVE

A complete analysis of the wobbling drive of gaseous media inside a cylindrical chamber is a rather difficult problem; however, to get a general idea, consider a two-dimensional problem with incompressible fluid under zero-order dynamic conditions (i.e. no higher order wave motion), and let the viscous shear be concentrated at the periphery of the boundary between fluid and container.

Figure 7.6 illustrates the mechanization of this simplified model. The container of the model is represented by a circle with a radius r_1. Three brackets attached to the container are guided by three small cranks with radius a. One of the cranks provides the driving torque at speed ω. Assuming that the fluid is also whirled around at speed ω by the wobbling motion of the container and that its volume is considerably less than that of the container, the fluid will be "plastered" against the inside wall of the container by the centrifugal force of the fluid forming a hollow core. However, the fluid also experiences a second mode of motion, since its mass center is being swung around with respect to the center of rotation, C_r. Now, since the fluid is assumed to be rotating inside the container at speed ω (the same as the

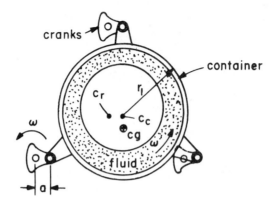

Fig. 7.6. Basic configuration of a wobbling fluid drive.

container), it follows that the configuration of the fluid, as defined by its boundary, remains unchanged with respect to these two centers.

To get a feel for the dominant parameters, it is postulated that:

- The pressure of the fluid against the container will always be normal to the container surface, so that the resultant pressure must pass through the center of the container.
- The revolving fluid has a self-centering property which tends to maintain the void at the core. Any lateral acceleration applied to the container will displace the void space and, at the same time, generate a pressure force on the surface of the container. This force will pass through the center of the container and the center of the void, which is diametrically opposite the center gravity (c.g.)
- The wobbling motion dictates that the fluid must rotate inside the cylinder against a shear moment. The moment must be balanced by inertial force originating from the c.g. of the fluid
- The resultant acceleration reaction force of the fluid must pass through the c.g. of the fluid and the center of rotation C_r.

Based upon the above hypothesis, a force equilibrium diagram may be constructed as shown in Figure 7.7 where:

f_c = centrifugal force with respect to the rotation center of the center of the container

f_s = centering force of the fluid with respect to the container

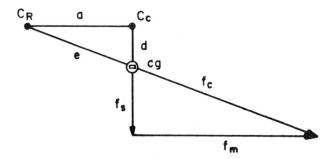

Fig. 7.7. Simplified force equilibrium diagram.

f_m = resultant force acting upon the c.g. of the fluid

$f_m d$ = inertia drive moment to balance against the shear between the fluid and the surface

To clarify the simplified model in Figure 7.6, a further analogy is shown in Figure 7.8. In this configuration, a mass mounted on a bar constrained by a spring is whirled by wobbling, but not rotating, a pin with center C_c that orbits around a circle with center C_r. In this manner, it is easy to see that the entire structure maintains an equilibrium with respect to C_r. The centrifugal force f_c would pass through C_r and the center of gravity of the mass at c.g., with one component f_s in the direction of the guiding bar in order to stretch the spring, and the second component f_m normal to the bar to provide the moment from acting against the friction between the pin and bearing.

The analogy between the schemes in Figure 7.6 and Figure 7.8 may be observed by considering that the container in Figure 7.6 serves the same function as the bearing of the swinging mass in Figure 7.8, except that its diameter is much larger than that of the bearing. Thus in Figure 7.8, the mass revolves outside of the bearing surface where the friction force is taking place, whereas in Figure 7.6, the fluid revolves inside the container. The natural tendency of the revolving fluid inside the container is to form a ring of uniform thickness and thereby to have the c.g. coincide with the center of the container. When the container and the fluid are subjected to lateral acceleration, the corresponding lateral force deforms the ring and shifts the c.g. away from the center of rotation. In essence, the revolving fluid ring has a certain "elastic stiffness" to resist the lateral force; the elastic stiffness of the revolving fluid is equivalent to that of the spring in Figure 7.8.

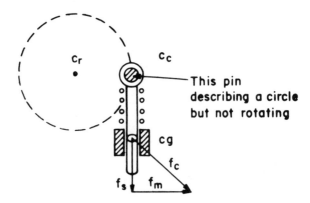

Fig. 7.8. Simple conceptual model of wobbling fluid drive with damped parameters.

Analysis of the Parameters of Figure 7.6
to Match that of Figure 7.8

Determination of the Equivalent Stiffness

In the "zero order" mode, i.e., no wave motion, the inside surface of the fluid is assumed to retain the shape of a circle, shown in Figure 7.9. When the c.g. is displaced
downward by a distance d, the center of the inside circle c_i would be displaced upward
by a distance of x in Figure 7.5. One way to compute the relationship between the
various parameters of the configuration of the revolving fluid is to introduce a hypothetical float occupying the space inside the fluid. The same displacement can be
achieved either by pushing the float upward or accelerating the whole structure upward with the same amount of force (assume the container has no mass). The inspiration of utilizing a fictitious float came from a party game where a young professor of
Chemistry from Cornell University asked if a balloon were placed inside a barrel and
both pushed sidewise, which way would the balloon go?

The force required to displace the float may be computed from the pressure gradient of the fluid as a result of rotation. Thus:

pressure gradient $= \omega^2 r \rho$

   r = radius

   ω = angular speed

   ρ = density

The reaction force per unit length of the float (Fig. 7.9) with a displacement of x normal to the circumference is:

$$df = \omega^2 r_2 \rho x.$$

The total force to displace the float would be:

$$f = 4r_2{}^2\rho x\int_0^{\frac{\pi}{2}} \sin^2\theta\, d\theta;$$

thus the equivalent stiffness is:

$$f/x = \pi\omega^2 r_2{}^2\rho;$$

where:

$r_2 d\theta$ converts angle θ to circumferential length;
$x\sin\theta$ converts common displacement x to the displacement normal to the surface;
the second $\sin\theta$ converts pressure force to a resultant force;
"4" represents 4 quadrants.

The displacement d of the c.g. from C_c, as shown in Figure 7.9, can be computed directly from the cross sectional area of the fluid when the hypothetical float is displaced by the distance x:

$$d = \frac{\pi r_2{}^2 x}{(r_1{}^2 - r_2{}^2)}$$

$$\frac{d}{x} = \frac{r_2{}^2}{r_1{}^2 - r_2{}^2} \tag{2}$$

Divide (1) by (2):

$$\frac{f}{d} = \pi\omega^2\rho(r_1{}^2 - r_2{}^2) \tag{3}$$

$$= m\omega^2,$$

where

$m = $ total mass.

The force f results from assuming there is a flotation force with the tendency to bring the c.g. back to the geometric center of the container. On the other hand, if we treat the fluid as a rigid body with a c.g. displaced a distaince d from the center of rotation, which is the center of the container of Figure 7.9, there should be a centrifugal force f_c pointing away from the center of rotation with a magnitude also of:

$$f_c = m\omega^2 d = -f. \tag{4}$$

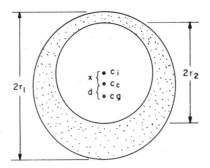

Fig. 7.9. Float concept of a void inside revolving fluid.

The above analysis indicates that the revolving fluid in the cylinder may assume a stable condition with a positive spring stiffness (equation 3), which also characterizes the spring in Figure 7.8, and thereby maintain the configuration of Figure 7.6. This stable condition is deducted according to postulation item 2 outlined before. On the other hand, the configuration in Figure 7.6 may assume an unstable mode according to equation 4. In the unstable mode, the fluid tends to lump together at the diametrically far end of the wobbling container. In this configuration, it would not serve the nuclear propulsion rocket function which requires a laminated flow pattern of the hydrogen and gaseous nuclear fuel. However, the author and his colleagues believed that, under certain conditions and over a period of time, it is possible to maintain the stable condition and force the gaseous nuclear fuel to whirl inside a chamber 1 foot in diameter up to a speed of Mach 1 with a small wobbling radius, such as .01 inch. During that period, the only regret was that there was no funding for nuclear propulsion.

In the following sections, other applications to wobbling drive will be discussed. In all the new applications, it has been experimentally proven to be desirable to have the fluid lumped together.

6. FAST EVAPORATOR FOR DESALINATION AND OTHER PROCESSES

Figure 7.10 depicts the essential parameters involved in a continuous evaporator. Fluid is admitted into the elementary block at the rate x and emitted at the rate of $x - y$, with y representing the evaporation rate. Energy balance is established as:

$$Ky = c(T_1 - T_2) \ \frac{ab}{h} = H, \tag{15}$$

where:

K = thermal capacity of the fluid being evaporated
c = thermal coefficient of conductivity
$(T_1 - T_2)$ = temperature gradient across the fluid layer
H = heat energy flow rate

The ratio of y/x for satisfactory operation is effected by the scale forming tendency on the evaporating surface, because the solid ingredient left behind by the y portion of the evaporated fluid must be carried away by the remainder portion of the fluid $(x - y)$.

In some evaporation systems, it may be desirable to introduce a reflux circulation parallel to the main throughput as shown in Figure 6.11. In concept, this added branch does not disturb the thermal balance of the main throughput, except that it will increase the velocity of the fluid passing over the evaporator system, as shown in equations 6 and 7 of Figure 7.11. The added velocity $V + \Delta V$ may be desirable to increase the thermal coefficient of conductivity C and the scrubbing effect for removing the scale.

According to equation 5, for a specified K, y, c, and $(T_1 - T_2)$ (with

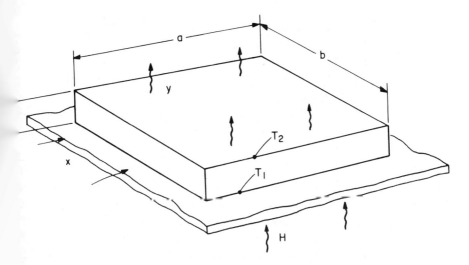

Fig. 7.10. Essential parameters of continuous evaporating surface.

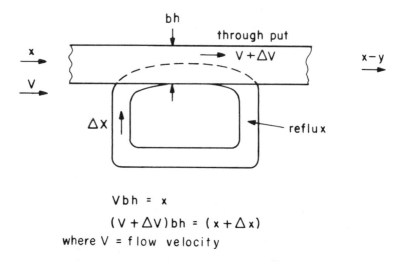

$$Vbh = x$$
$$(V + \Delta V)bh = (x + \Delta x)$$
where V = flow velocity

Fig. 7.11. Flow pattern including reflux.

or without the use of reflux), the product $a \times b$ is therefore inversely proportional to h, where $a \times b$ represent the physical size of the evaporator.

How to minimize h and, at the same time, increase V are therefore two key parameters of concern when searching for a better configuration for the evaporator of which the desalination system is but one special case. In the following analysis, we will see how wobbling drive might be an ideal configuration for this purpose.

Figure 7.12 shows the scheme of a modern evaporator desalination system, where heat energy needed to evaporate a film of sea water on one side of the heat exchanger wall is supplied by heat recovered from condensation of the vapor at the other side of the same wall. A vapor pump is used to maintain a pressure gradient on the two sides of the wall to establish the equilibrium needed for evaporation and condensation, as the case may be.

The scheme to be introduced here involves a group of tubes with the fluid to be evaporated flowing inside. Vapor will condense on their outside surfaces, while energy released due to vapor condensation will supply the energy needed for evaporation.

The tubes are driven to form a wobbling motion, so that the main flow stream of the fluid will be thrown by centrifugal force against the far side of the tube from the center of the wobbling, as shown in

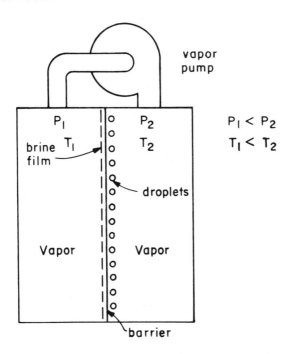

Fig. 7.12. Modern desalination concept (vapor recompression).

Figure 7.13. The main flow stream forms a crescent-shaped cross section which is forced to revolve inside the tube.

Since the tube wobbles but does not turn, it follows that the crescent-shaped water column will be forced to revolve with respect to the tube, thereby coating the inside surface of the tube with a thin film. The thickness of the film is a function of the revolving speed and the viscosity of the fluid. For water, it is a relatively simple matter to coat the inside wall with a very thin film of .001 inch to .005 inch.

Application to Desalination Process

Figure 7.14 shows a desalination system utilizing an evaporator designed from the concept of wobbling tubes. The system is enclosed in a metal case (1); inside metal case (1), there are pairs of evaporator chambers to keep the dynamic force in balance. In this figure, two chambers (2) and (2′) are shown. Groups of tubing identified as (3)

reference
point on
tube

fluid

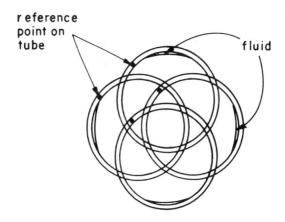

Fig. 7.13. Revolving fluid in a wobbling tube.

are secured to the end walls (27) and (28) of chamber (2). Chamber (2) is mounted on at least three cranks, two of which are shown in this diagram. The first crank is shown as (4); it has a pivot (5) secured with chamber (2) and a shaft (6) secured at the bottom of the metal case (1). This crank also provides the drive by having a gear (7) coupled to the shaft (6). This is then driven by pinion (8) coupled to motor (9), as shown in the diagram. Chamber (2') is driven by crank (4') through shaft (6') and gear (7') and coupled to the same motor driven by pinion (8). Crank (5') is oriented 180 degrees out-of-phase with crank (5) so that in operation, the oscillating force generated by the chamber (2) is exactly balanced by that of chamber (2'). For chamber (2), the second set of cranks is shown as (10, 11, and 12). This particular crank is hollow inside and also used as the outlet for the condensed fluid. Chamber (2) has an inlet (17) coupled to the stationary case (1) through a flexible conduit (18, 19, 21, and 20).

For desalination purposes, brine is introduced into the system by pipe (15) and distributed by rotating tubes (14) which leads the brine inside tube (3). The stream of brine flowing down tube (3) is identified as (26). Excess brine flows to the lower part of chamber (1) and drains through pipe (16). The vapor formed by the evaporation process is sucked out by the centrifugal fan (22). The inside space of the case (1), identified as (23), would normally be filled with vapor. The fan (22) creates a pressure difference between its outlet (24) and inlet (23). From outlet (24), the vapor with higher pressure is forced

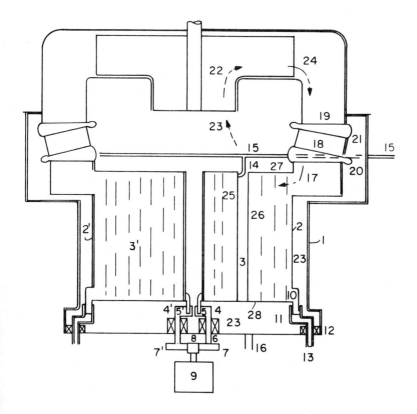

Fig. 7.14. Vapor recondensing desalination system with wobbling drive.

into chamber (2) via the flexible conduits (19, 21, 18, 20 and 17) and
envelops the outside surfaces of tubes (3). Being higher in pressure,
the temperature of the vapor outside of tubes (3) will also be warmer
than the inside and will condense on the tube to form droplets (25).
The heat energy thus released is transmitted across the wall of the tube
to evaporate the fluid (26) coated on the inside surface of the tubes
(3). The condensation (25) outside of tube (3) is collected at the bottom of chamber (2) and drains from tube (13). Figure 7.15 shows the
top view of the fluid distribution system which brings the water from
pipe (15) through a revolving joint (30). The revolving joint (30) coin-
cides with the wobbling center of tube (3) so that (14) and tube (3) revolve together. A slotted opening (29) on revolving tube (14) distrib-
utes the fluid to the far side of the inside surface of tube (3).

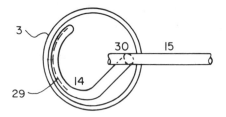

Fig. 7.15. Inlet distribution for the wobbling tube.

On casual inspection, the system depicted in Figure 7.13 is similar to an ordinary desalination system with stationary, vertically mounted, heat exchanging tubes. The added complication of wobbling drive and the flexible conduit may seem to be unwarranted. However, to be more specific, this system provides drive for the fluid along the circumferential direction over the inside surfaces of the wobbling tubes (3) of Figure 7.13. The circumferential direction of the flow of the fluid corresponds to the flow rate x shown in Figure 7.10, whereas the apparent flow direction of the fluid along the length of the tubes is essentially the distribution system bringing the fluid to the head station of all primary drive systems in the circumferential direction.

The fluid of an evaporator cannot be driven by a conventional pressure head, because the open surface needed for evaporation would not be able to maintain a pressure gradient. As a result, the primary flow path of the evaporator must be moved under the influence of gravity or acceleration.

In addition to the driving force along the direction of flow, there should be a stabilizing effect to keep the fluid film distributed uniformly along its width. In principle, this objective can be achieved by a uniform gravitational field across the path. In the case of bringing a fluid down the side of a vertical surface, the lateral gravitational field is zero, and hence there is no stabilizing tendency to keep the stream uniformly thick.

One scheme now in use places evaporating tubes in a horizontal position to let the fluid fall in sheets onto the tubes, wrapping themselves around the tubes as shown in Figure 7.16. At the stagnation point a, or leading edge of the tube in this arrangement, the sharp turn of the fluid path creates a high lateral, acceleration gradient which

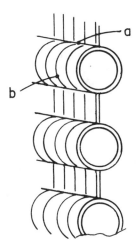

Fig. 7.16. Generation of water film over horizontal tubes.

may therefore produce a strong smoothing effect for maintaining uniformity in the fluid film thickness. However, as the fluid moves away from the leading edge into area b, the curvature of the tube tends to produce a negative gravitational effect (pulling away from the surface) which can be unstable with regard to film thickness.

In the scheme herein proposed, the fluid film is subject to a positive gravitational effect at all times, so that the uniformity of film thickness is assured. Viewed in this manner, wobbling motion is in fact effective in most efficiently driving the fluid in the desired manner.

Figure 7.17 illustrates the observed flow pattern of a revolving water stream inside a transparent wobbling tube under the illumination of a strobe light. The stream behaves like a wet mop wiping the inside surface of the tube to agitate the evaporation process. This particular innovation and exercise were stimulated by the following:

- wobbling drive for gaseous, nuclear rocket propulsion;
- the almost invisible film left behind the receding fluid column when operating the Fluidyne described in the previous chapter (the rapid evaporation of the film provides the driving action of the Fluidyne);
- the author's knowledge of the desalination process developed by Israel Desalination Engineering Ltd.

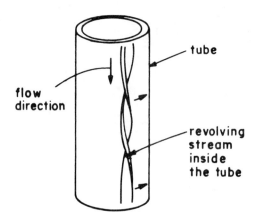

Fig. 7.17. Revolving stream observed under strobe light.

Evolution of the wobbling drive distillation scheme from these three separate events would not have occurred had the author not made a parameter analysis of each system and recognized the need for thin film and bypass flow in the case of the desalination system and the existence of these parameters in the two other configurations. The university teaching environment seems to foster cross-pollination of interesting ideas.

The practical adaptation of wobbling drive for desalination seems to be quite good. The Israeli system is rated at 15 KWh per cubic meter of distilled water compared to 12 kwh for the reverse osmosis system however, the reverse osmosis system is difficult to service and not quite ready for the general market.

The advantages of wobbling drive versus the Israeli scheme (Figure 7.17) are:

- thinner fluid film more evenly distributed and hence more efficient;
- evaporation taking place inside the tube facilitating the descaling operation service.

The disadvantage lies in the complicated nature of the wobbling drive and determining the exact cost effectiveness of the new scheme requires some further experimentation and study.

7. A CONCEPTUAL STUDY OF A CONTINUOUS DISTILLATION COLUMN

Figure 7.18 illustrates the ingredient enrichment process which plays the key role in a distillation system. In this diagram, the liquid and vapor phases of the solution of two ingredients, such as water and alcohol, or benzine and toluene, are positioned to achieve a counter flow pattern with the liquid traveling toward the left and the vapor to the right. Across the vapor/liquid interface, an inequilibrium condition is maintained by manipulating partial pressure and temperature of ingredients, so that molecules of ingredient I, represented by ↑, tend to move from the liquid to the vapor, thus enriching the content of I in the vapor while it is flowing to the right. On the other hand, the molecules of ingredient II represented by ↓ tend to enrich it in the liquid flowing toward the left. A complete distillation system is essentially an extension of the counter flow arrangement of Figure 7.18, which achieves a desired degree of enrichment.

A thorough study of a distillation system requires knowledge of heat, mass, and momentum transfer technology, chemical engineering, and the specific configuration of the distillation tower. However, from the simple diagram in Figure 7.18, one can appreciate that the dominating parameters of a distillation system are:

- a large interface area between fluid and vapor per unit volume of the equipment;

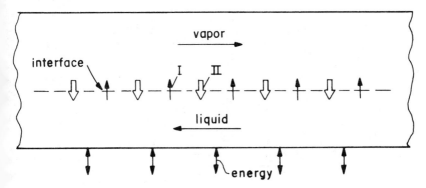

Fig. 7.18. Basic distillation process.

- the means of maintaining a high inequilibrium condition, herein called the L factor, to promote molecular transfer across the interface;
- the means of manipulating energy flow to support the desired L factor at each station;
- the means of maintaining the counterflow pattern with a flow rate in balance with the interface area.

The conventional distillation tower is an ingenious scheme, utilizing the "bubbling" effect of vapor forced to emerge through tin, perforated holes at the bottom of trays containing the fluid, in order to create a large interface area between vapor and liquid. In a distillation tower, hundreds of these trays are stacked together, as shown in Figure 7.19, where the dotted line arrow represents the upward flow pattern of the vapor and the solid line the downward and zig-zagged flow pattern of the liquid.

Since a pressure gradient is needed to force the vapor through the tray, there is a sizeable, total pressure drop from the bottom to the top of the tower; in general this is designed to match the energy requirement of the operation. Such a system uses mechanical energy to supply the needed thermal energy, which is inefficient energy utilization.

As an exercise in innovation, the adaptation of wobbling drive to distillation is tempting. The configuration is quite similar to that in Figure 7.14, with the exception that the wobbling tubing is used to carry fluid and to generate the interface surface between fluid and vapor which flow inside the fluid passage.

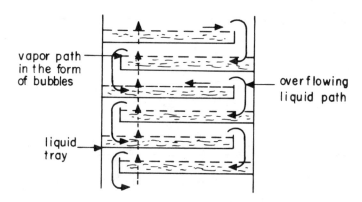

Fig. 7.19. Bubbling plate distillation tower.

One apparent drawback of wobbling drive in comparison to the "bubbling trays" is the relatively smaller interface surface area between fluid and vapor per unit volume of equipment. However, in its favor is the chance of increasing the inequilibrium factor L over that of the bubbling tray system by the free, turbulent flow pattern of vapor inside the thin, liquid film lining the tubes of the wobbling drive system. The following explanantion may be given:

- In the hollow core of the fluid tube of the wobbling drive, molecule movement across the interface area is accelerated by the free circulation of vapor over the interface area. Conversely the bubble of the conventional distillation tower traps the vapor and tends to limit its movement.
  As an analogy, foamed plastic material is used as a heat insulator because of the reduced molecular activity of air trapped inside the foam compared to the open conduit in wobbling drive.
- The froth of liquid in the bubbling tray is a poor heat conductor, so that the energy balance needed to support the molecular exchange must be stored in the specific heat of the fluid in the bubble, which is not an efficient way of maintaining a continuous operation.
- Because of these two conditions, a conventional distillation tower operates on the multiple-equilibrium-stages principle, which matches well with the multi-plate (or tray) distillation tower configuration. In designing the tower, the goal is usually to achieve a "ladder" of equilibrium conditions along the liquid-vapor equilibrium diagram of a particular solution (Figure 7.20). In so doing, the operation tends to linger near equilibrium, instead of developing a large inequilibrium factor L which is responsible for enrichment.

8. PARAMETER ANALYSIS OF A CONTINUOUS DISTILLATION PROCESS: PART I, EQUILIBRIUM AND INEQUILIBRIUM

The objective of the preceding analysis was to link a few key parameters of a continuous distillation process in order to achieve an optimum condition. This exercise was carried out as an example for the invention class at M.I.T. to illustrate the parameter analysis process; experts

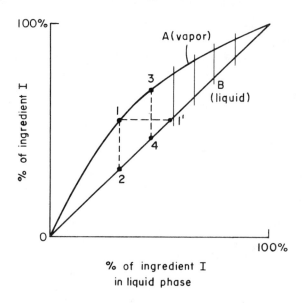

Fig. 7.20. Equilibrium diagram of liquids in solution.

in chemical engineering may find this analysis unconventional, naive, or even erroneous. The author can offer no defense except to apologize for having very little prior experience in the field; of course, this may also illustrate the spirit of innovation, because, after having identified the key parameters, one should not hesitate to plunge into a new area: Thoughts can only be polished after making a few mistakes through several stages of trial formulation.

Figure 7.20 represents a typical equilibrium condition diagram of two ingredients completely soluble in each other, with a mixture ratio defined as the percentage (from 0-100%) of one ingredient (or the other) in the mixture. The two curves A and B indicate that for a given temperature and pressure, there is one mixture ratio represented by the percentage of ingredient I in the vapor phase (point 1) coexisting in equilibrium with a second mixture ratio in the liquid (point 2). This phenomenon may be described as the difference in the partial pressure of the vapor of the two ingredients.

For years, people have realized that ingredients can be separated from a solution by utilizing a successive distillation process. Conceptually, this is illustrated in steps 1, 2, 3, 4 of Figure 7.20. Liquid phase station 2 is in equilibrium with the vapor phase at station 1, which is richer in ingredient I than in station 2. Mechanically, it is possible to separate a portion of the vapor at station 1 from that at station 2 because of their differences in density.

The vapor at station 1, after being separated from the liquid at station 2, may be totally condensed into liquid as represented by $1'$, which has the same ingredient ratio as station 1; however, if a portion of it is allowed to evaporate, a new vapor-liquid state equilibrium will be established as represented by stations 3 and 4. Thus, it is possible to accomplish the total separation of two ingredients by the repetitive use of mechanical separation and by re-establishing equilibrium conditions.

In the step-by-step operation described in the last paragraph, there is a progressive diminution in the total amount of vapor being drawn out for the next stage, while ingredient I is progressively enriched. Remaining after each stage is the liquid residue, also in a ladder of concentration of ingredient I, but lagging, in terms of its concentration, behind that of the vapor ladder. This liquid ladder can be reused by combining it with the supply of similar concentration in the batch to follow. Even with this provision, a pyramid seems to exist, in that the size of the successive stages is reduced while supporting a steady rate of output with the desired high concentration of ingredient I. Conceptually, this situation may be illustrated as Figure 7.21.

The use of reflux to achieve continuous and uniform flow for distillation was a major innovation. The logic was probably established by reasoning that if a pyramid pattern is needed to concentrate ingredient I, a reversed pyramid pattern would be needed to concentrate ingredient II in the opposite direction. This would mean that a quantity of the solution with high concentration of ingredient I must be used to start the reverse process, which seems illogical because concentration of ingredient I was the primary objective. However, a continuous uniform flow system is conceptually elegant and can be achieved by matching the two, opposite flow patterns. In so doing, a fairly large amount of reflux must circulate within the entire system from one end, where there is a high concentration of ingredient I, to the other end, where there is a high concentration of ingredient II. By circulating the reflux, the input is added at some mid-station, while the desired high concentration of ingredient I can be tapped from the top and the residue with high concentration in ingredient II can be removed from the other end. The advantage of a continuous system lies in the

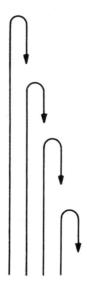

Fig. 7.21. Diminishing return in distillation batch process.

automatic balance of the energy needed at each state, as well as uniformity in handling the total flow.

9. PARAMETER ANALYSIS OF
A CONTINUOUS DISTILLATION PROCESS:
PART 2, THE CONCEPT OF OPTIMUM DESIGN

In the last section, the need for a large amount of reflux for continuous distillation was established and maintaining a large L factor across the interface was also recognized. In this section, an attempt will be made to fit these requirements together to attain an optimum design.

The flow pattern of a continuous distillation system may be laid out as shown in Figure 7.22, where the heavy dashed line represents the interface, and the various shaded areas represent the different ingredients; arrows represent the direction of flow and transfusion of ingredients. Indeed, there are simple distillation devices in the shape of slender tubes, packed or unpacked, which operate on the general principle described above. As of now, however, the author does not know whether analysis of the slender tube distillation device is handled by

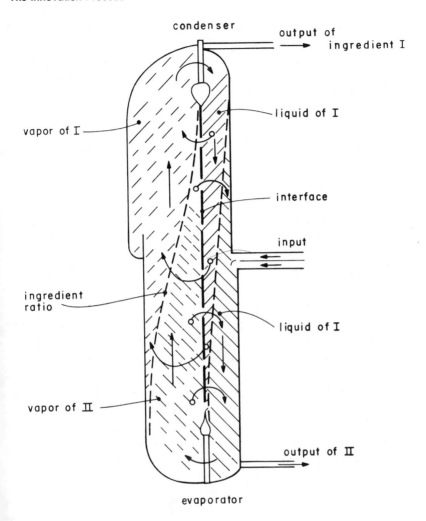

Fig. 7.22. Conceptual layout of a continuous distillation column.

the finite number of plates method or by a similar method described next.

The next step in making the analysis is constructing a general flow pattern diagram with flow rates identified as shown in Figure 7.23. The symbols used to analyze the mass flow rate are summarized as follows:

1. Input is represented by a flow rate of $1/1$. The number to the upper left of the slanted line represents the flow rate of ingredient I, while the number to the lower right represents the flow rate of ingredient II.

2. Outputs are represented by $1/0$ at the top for high concentration in ingredient I and $0/1$ at the bottom for high concentration in ingredient II.

3. For symmetrical operation, input is split into two $.5/.5$ branches: One branch is to be evaporated and added to the vapor branch of the column, while the other is to be added to the liquid branch.

4. The column is represented by a vapor branch rising on the left side of the column and a liquid branch flowing downward on the right side of the column; the two branches are separated by an interface represented by a dash-dot line.

5. The column is provided with a primary reflux flow $A/$ and $/B$ and two, secondary reflux flows $x/$ and $/y$; they may be considered the carrier vehicles, because it is the reflux media $A/$, $/B$, $x/$, and $/y$, which flow continuously like conveyor belts carrying ingredients I/I around, thereby performing the separation.

6. The consistency of the flow pattern in the two ingredients may be checked separately as follows:

   • To check ingredient I, use station 1 in the liquid column immediately before admitting the input as the starting point; the starting flow rate is $A/$.

   • The flow rate becomes $(A + .5)/$ at station 2 after the influx of half the input. The entire flow rate of $(A + .5)/$ makes a transfusion across the interface between station 3 and station 6, thereby reducing the flow of ingredient I to zero between stations 2 and 5. In this manner, the output at the bottom of the diagram is free of ingredient I.

   • Between stations 7 and 8, the flow rate of ingredient I, now

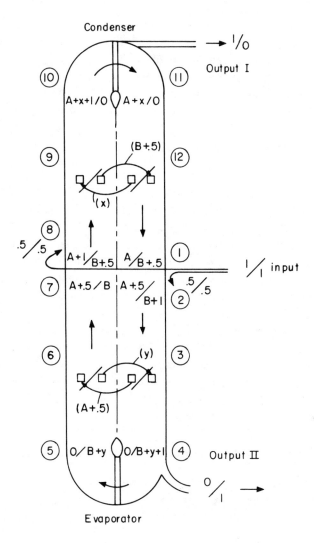

Fig. 7.23. Conceptual flow pattern diagram of a distillation system.

in vapor form, changes to $(A + 1)/$, due to the input of .5/, also in vapor form. At station 9 across the interface with station 12, there is a transfusion of $x/$. This flow rate $x/$ may be considered a secondary reflux localized in the upper portion of the distillation column. Between stations 10 and 11, the entire flow rate $(A + x\ 1)/$ of vapor passes through a condensor to become liquid. After extracting the output with a flow rate of 1/, the flow rate of ingredient I passes stations 11, 12 and back to 1 to resume the initial flow rate of A.

The flow pattern of ingredient II may be traced in like manner, starting from station 7 with an initial value /B. At station 8, this flow is added by the vapor input to become /$(B + .5)$ which is then transfused across the interface of the upper half of the column, as represented by stations 9 and 12, by-passing stations 10 and 11. In so doing, the output at state 11 is free of ingredient II. After flowing down through stations 1 and 2, ingredient II becomes /$(B + 1)$ when entering station 3. Thereafter, it is added across the interface by the secondary reflux /y to become /$(B + y + 1)$. At this lower end of the column, the output flow rate /1 of ingredient II is removed with the remainder, /$(B + y)$ being evaporated through the evaporator. Since the secondary reflux /y is limited to circulating around the lower half of the column, the flow of ingredient II returns to station 7 with its initial value /B.

The Optimum Value of the Primary Reflux Flows A and B

One can assume $A = B$ and $x = y$ in the following discussion for simplicity's sake. This is particularly applicable to those mixtures with symmetrical properties like those exhibited in a benzene-toluene solution.

At the point of interface between station 8 and station 1, we have, for the vapor phase, a mixture strength for ingredient I of:

$$\frac{A + 1}{A + 1 + B + .5} = \frac{A + 1}{2A + 1.5} \tag{1}$$

whereas for the liquid phase, a mixture strength for ingredient I of:

$$\frac{A}{A + B + .5} = \frac{A}{2A + .5}. \tag{2}$$

The relative strength of ingredient I vapor phase versus the liquid phase is tabulated in Table 7.1:

Table 7.1

a	A	1	2	3	4
b	Strength of I in vapor	.57	.55	.53	.526
c	Strength of I in liquid	.4	.44	.46	.47
d	Difference $= b-c$	.17	.11	.07	.056
e	Non-equilibrium $L = .2-d*$	.03	.09	.13	.144
f	$\dfrac{L}{B + .5} = \dfrac{L}{A + .5}$	.02	.036	.037	.032

In Table 7.1, line a shows four assumed values of A or B, varying from 1 to 4, which represent the ratio of the primary reflux with respect to the feed, identified as unity. Lines b and c show the strength of ingredient I in vapor and in liquid across the interface between stations 8 and 1. Line d shows the difference between the values of lines b and c. Using benzene and toluene as a reference, as shown in Figure 7.20, the equilibrium condition for a half-and-half solution is about .20.* Thus, any deviation from this value for the half/half solution would represent the L factor or the rate of diffusion of the ingredients across the interface. The values of the L factor shown in line e are therefore a function of A, shown in line a. It is further assumed that the rate of diffusion across the entire interface is to be adjusted later in proportion to the rate chosen at the middle section. Thus, an optimum value of reflux A or B exists, because, while the L factor is increased by the value of reflux shown in line e, the reflux itself (which acts like a burden) must be transfused by the L factor. This trade-off is expressed by the ratio of the L factor to the total amount of ingredient II transfused across the interface of the upper half of the column, as shown in line f. According to line f of Table 7.1, the optimum primary reflux appears to be around $A = B = 3$.

The Optimum Value of the Secondary Reflux X and Y

As shown in Figure 7.23, the values of the secondary reflux flow rates x and y are localized at the two ends of the column. Their values

---

* .2 is the deviation of the equilibrium condition at the center of the vapor phase curve shown in Figure 6.19.

are to be determined in order to match with the given, primary reflux flow rate and in order to optimize the transfusion rate.

The search for the optimum value of x and y is carred out according to the following plan:

- Value of A and B is assumed to be 3 (three times the output rate), as determined by earlier optimization (Table 7.1). This may be considered the result of the first round. To do a complete optimization, several iterations may be needed.
- The total amount of the primary reflux of 2 and the load of .5 is divided into five equal parts to be transfused across the interface at five sections, identified as n, which varies from 0 to 5.
- The amount of x is to be chosen to optimize the L factor and thereby maximize the transfusion rate of primary reflux, output, and x itself.

For this computation, x may be selected independently of A or B, because, according to the mass flow diagram in Figure 7.23, x or y may be independently adjusted without violating the law of conservation of mass.

Allowance must be made, however, for supplying local energy needed to facilitate the transfusion. For instance, if x is assumed to be zero, then energy must be taken out across stations 9 and 12 of Figure 7.23 to condense ingredient $(B + 5)$ from the vapor phase in station 9 to the liquid phase in station 12.

- For all trial values of x, the transfusion across the interface is assumed to be uniformly distributed from section n = 0 to n = 5.

Table 7.2 represents a sample calculation of three sets of distribution of x, for x = 1, 2, and 3. Referring to Figure 7.23, section n = 5 represents the top part of the column across stations 10 and 11, where the value of x is at its maximum. Section n = 0 is near the center of the column, where x = 0. Column a of Table 7.2 shows the value of the two ingredients in the vapor phase, whereas column b shows the value of the two ingredients in the corresponding liquid phase. Column c shows the difference in the consistency of the vapor over the liquid in terms of the ratio of ingredient I.

The actual number of various values of x is shown in column d; column e represents the difference in the consistency of the vapor over the liquid in a corresponding equilibrium condition. This is determined

Table 7.2

η	a Vapor	b Liquid	c Consistency difference of Ingredient I	x = 1 d	x = 1 e	x = 1 f	x = 2 d	x = 2 e	x = 2 f	x = 3 d	x = 3 e	x = 3 f
5	$4+\frac{x}{0}$	$3+\frac{x}{0}$	$\left(\frac{4}{4}+x\right)-\left(\frac{3}{3}+x\right)$	0	0	0	0			0		
4	$4+\frac{.8x}{.7}$	$3+\frac{.8x}{.7}$	$\left(\frac{4}{4.7}+.8x\right)-\left(\frac{3}{3.7}+.8x\right)$	.872 .844 .028	.095	.067	.888 .867 .021	.081	.060	.901 .885 .016	.070	.054
3	$4+\frac{.6x}{1.4}$	$3+\frac{.6x}{1.4}$	$\left(\frac{4}{5.4}+.6x\right)-\left(\frac{3}{4.4}+.6x\right)$	.767 .720 .047	.150	.103	.787 .750 .037	.138	.101	.805 .774 .031	.123	.092
2	$4+\frac{.4x}{2.1}$	$3+\frac{.4x}{2.1}$	$\left(\frac{4}{6.1}+.4x\right)-\left(\frac{3}{5.1}+.4x\right)$	.680 .620 .060	.180	.120	.695 .644 .051	.175	.124	.712 .666 .046	.165	.119
1	$4+\frac{.2x}{2.8}$	$3+\frac{.2x}{2.8}$	$\left(\frac{4}{6.8}+.2x\right)-\left(\frac{3}{5.8}+.2x\right)$	.600 .530 .070	.200	.130	.611 .548 .063	.200	.133	.621 .562 .059	.189	.130
0	$4+\frac{0}{3.5}$	$4+\frac{0}{3.5}$	$\frac{4}{7.5}-\frac{3}{6.5}$	.533 .401 .072	.200	.128	.533 .461 .072	.200	.128	.533 .461 .072	.200	.128

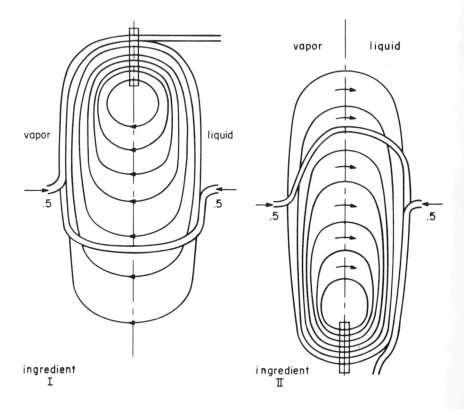

Fig. 7.24. Distribution pattern of ingredients in a distillation column.

from the equilibrium diagram of Figure 7.20 by reading off the consistency of the liquid phase; for example, at n = 4, $x = 1$, $e = .095$, which is the distance between the two lines at .844 on the horizontal axis of Figure 7.20. Column f represents the L factor which is the difference of $e - d$.

Figure 7.24 shows the flow distribution of the ingredients I and II plotted separately using $A = B = x = Y = 3$. From this diagram, the flow path of the two ingredients, moving from input to output, is quite clear.

10. REFLECTION

"Heat, mass momentum transfer," as an engineering discipline usually deals with the flow of one particular physical quantity at a time. In teaching, the three are lumped together as one discipline, because of the similarity of their physical properties, especially when these properties are expressed in differential equations.

In the wobbling drive distillation system or the desalination system, all three physical quantities, heat, mass and momentum, are involved in an orthogonal manner. This situation alone is a fascinating one to study, but the important issue is recognizing the key parameters of the need, the general principle underlying the technology, and the configuration drawn from the bag of tricks, so that the final configuration materializes to satisfy the need.

In developing skill in innovation, one must be alert and make a point of examining key issues whenever confronted with a new system. It would be a good exercise to take students on a few plant visits and ask them to write down the key issues they observed. At the time the author visited the desalination plant in Israel, his input "information channel" was essentially jammed by exposure to a number of things which were totally new to him. After taking a moment to collect his thoughts, he began identifying the key issues and asking pertinent questions, which made the remainder of his visit most enjoyable.

The main purpose of this discussion is to set up a pattern for developing a teaching format which will drill students in thinking and analyzing in an innovative manner. If this material is used for teaching, the best approach is to lead the student step by step through a series of questions and answers with the hope that some new configuration may be generated.

8
The INTERFACE BETWEEN NONRIGID STRUCTURES AND GEOMETRIC CONFIGURATIONS

1. INTRODUCTION

In discussing the radial tire in Chapter II, we examined the combination of rubber, nylon, steel wires, and steel rims in various configurations that achieved a certain level of performance for automobile tires. In designing skyscrapers, super domes, and bridges with mile-long spans, high tensile strength cables, in combination with steel girders and concrete structure, became the rule instead of the exception. In airplane and rocket design, and even in sporting goods, such as golf clubs and skis, composite material incorporating exotic fibers has earned confidence and gained popularity.

All these engineering feats involve the classic principles of engineering science, such as the Lagrange equation, Timoshenko's Theory of Elasticity, and, more recently, the Finite Element Principle. These are powerful tools when an exact evaluation of the stress-strain relationship of certain basic configurations is desired, but it is difficult to see how product innovation can originate from the Lagrange equation or the Theory of Elasticity.

Kinematics is another branch of technology that parallels structure engineering. As a rule, kinematic structures involve rigid body members coupled together by flexible joints to achieve desirable shape, motion, and stress distribution. But, as in the examples of the radial tire, gripnail, and wire wrapper discussed in Chapter II, innovation often flourishes at the juncture of nonrigid structure and flexible geometric configuration. A few examples will be cited here on this interesting subject.

2. THE PROBLEM OF RESTORING THE TOWER OF PISA

Background

One of the oldest, nonrigid, man-made monuments is the Tower of Pisa, which began to lean eight hundred years ago, immediately after the lower section was completed. The idea of organizing an international team to tackle this challenging engineering problem was first suggested to Y.T. Li in 1965 by the late Professor Colonnetti who, at that time, was chairman of the committee for the restoration of the Tower of Pisa.

In response to an international contest inviting submission of proposals for restoring the Tower (announced in 1972 by the Department of Public Works of the Italian government), Li organized a technical advisory team in conjunction with the Vappi Construction Company, Cambridge, Massachusetts and put together the only proposal submitted from the U.S., thereby entering the competition with 15 leading construction firms from all over the world. The restoration program was postponed subsequently, partly because the rate of incline had not been accelerating significantly over the last few years.

From an informal source, it was learned that the Li-Vappi proposal was among five which reached the final stage of consideration. Furthermore, four of these five utilized the concept of soil stabilization where certain chemical ingredients are injected into the earth to increase its bearing strength. The drawback of the scheme was the unavoidable disturbance to the pressure bearing soil, now nearing the failure mode, during the injection process.

The Li-Vappi proposal was reviewed by Professor Travaglini, Commissioner of Public Works, with Y.T. Li in Rome on October 31, 1975. The discussion focused on the coupling between the tower base and the load transferring structure; Travaglini pointed out that after several hundred years of exposure, the stonework of the base had been weakened to the extent that it might not be able to take the 320-lb/in.^2

pressure applied to the base through the load transferring operation. A modified plan was submitted by Li on December 15, 1975, illustrating that it was possible to accomplish the load transferring process needed to maintain the overall stability of the tower without exceeding the existing stress level on the stonework of the base. The method proposed recognized the elastic property of various parts of the structure and took full advantage of that characteristic by programming the load transfer process accordingly. The logic behind this proposal follows.

The General Condition of the Tower

In designing the restoration system for the Tower of Pisa, the task had been made feasible by the availability of three excellent volumes, entitled *Ricerche e studi su la torre pendente di Pisa e i fenomeni connessi alle condizioni d'ambiente*, prepared by the 1965 commission for the Leaning Tower of Pisa. From these volumes, we have a complete picture of the Tower's physical structure and geo-technical information on the ground beneath the Tower. Some highlights of the structural information include:

- Tower weight   14,453 tons
- leaning angle   5°19'
- c.g. offset   2.24 meters
- tilt moment   32,400 meter/ton
- Tower base diameter   20 meters
- Tower base below the
  floor of the moat   2.88 meters

Geo-technical boring found no hard bearing strata down to 60 meters; they revealed layers of clay and sand. Of interest is the clay strata at a 10-meter depth from the ground surface, which shows considerable *dishing* under the influence of the Tower's weight, as shown in Figure 8.1

A record of the Tower's rate of incline has been compiled since 1911. This is graphed as a function of time, as shown in Figure 8.2, and exhibits considerable fluctuation from year to year. There seems to be some correlation between rate of incline and ground water level, with seasonal variations indicating a slower leaning rate during sum-

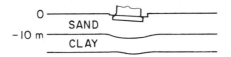

Fig. 8.1. Profile of existing soil foundation.

mer months when the water level is high. The recent addition of arte-
sian wells in the vicinity of the Tower caused considerable concern for
the Tower's safety. However, within the past 60 years, there does not
appear to have been a substantial increase in the average leaning rate,
which corresponds to 1 degree in 600 years. This means that the
Tower must have gone through a very rapid settling process in the
first 100 or 200 years to achieve the present 5°19' tilt. The continued
tilt increases the offset of the Tower's center of gravity and produces a
corresponding bearing stress on the south side of the Tower, which
may soon exceed the limited burden that the foundation can withstand.
As the maximum level of bearing pressure is reached, the rate of tilt
will accelerate until the Tower collapses. Since it is not possible to determine when this may occur, there is the general fear that any small
disturbance introduced by construction might trigger disaster. What
constitutes a tolerable level of disturbance is a matter of judgment.

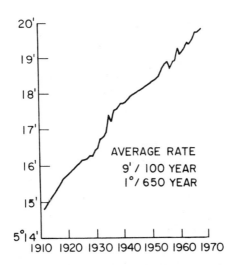

Fig. 8.2. Past record of the leaning angle.

Basic Design Configuration

The general design concept involves using a post tension concrete ring to couple a steel frame to the outside circumference of the tower base, as shown in Figures 8.3, 8.4, and 8.5. A pair of load distribution pads are then used to redistribute the load according to the tilt of the Tower, thereby preventing further incline. The coffer dam, shown in Figure 8.4 and the temporary load-carrying-boom of Figure 8.3 are safety measures and represent good engineering practice. In order to minimize the disturbance, it was decided that the post tension concrete ring should not exceed the 4-meter thickness of the tower base.

In the original design proposal, the post tension shear ring shown in Figure 8.5 was used primarily to transmit the shear loading from the surface of the tower base to the octagonal steel frame. However, as a result of the offset between the shear surface and the center line of the girder of the steel frame a twisting moment exists. This twisting moment is translated into a bearing pressure applied to the surface of the tower base. Since this added pressure is less than 240 ton/m^2 or 320 lb/in.^{2}, it is well within the loading permissible in ordinary engineering practice for masonry foundations. This value, however, is of particular concern to Professor Travaglini, as mentioned earlier.

The Modified Plan with a Load Transmitting Program

In the modified plan, it is believed that Professor Travaglini's concern can be alleviated by transferring excess soil bearing pressure to the steel frame through the shear loading of the stone base, while complementing the shear loading with an appropriate pressure loading, so that the equivalent internal stress in the stonework does not exceed the existing stress. Thus excessive pressure against the soil is relieved, but stress in the stonework is not increased. This appears to be the most achievable by any restoration scheme (including solidifying the ground), except maybe rebuilding the structure with new foundation.

The general procedure for reconstruction is to cast the ring in place but allow a gap between ring and base, as shown in Figure 8.6. As the cables are tensioned to bring the structure members into proper loading position, the diameter of the concrete ring will be compressed and the gap between ring and base reduced. With the concrete ring thus compressed, the gap will then be grouted to bond the ring to the base of

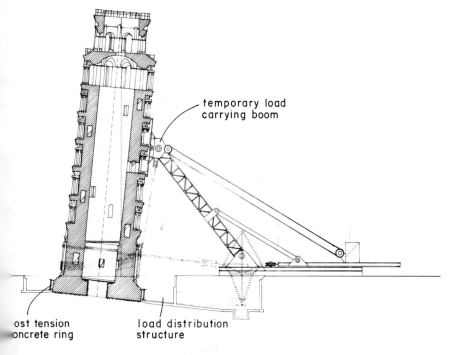

temporary load
carrying boom

ost tension
oncrete ring

load distribution
structure

Fig. 8.3. Proposed restoration scheme.

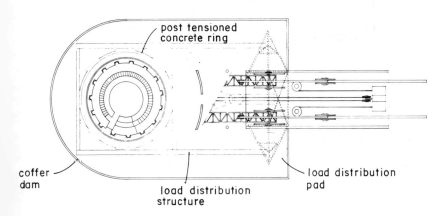

post tensioned
concrete ring

coffer
dam

load distribution
structure

load distribution
pad

Fig. 8.4. Top view of proposed scheme.

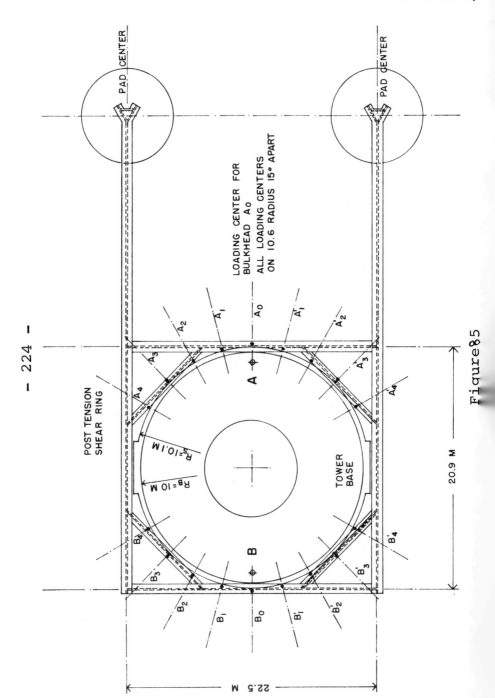

Figure 85

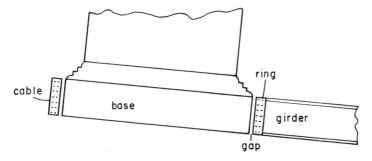

Fig. 8.6. Modified structure with a gap at the shear joint.

the tower. Any further tensioning of the cable will introduce pressure to its base, if so desired. However, there is an inconsistency concerning the gap; i.e., while the gap should be provided to allow the concrete ring to shrink in order to take up the load without compressing the base, there should be no gap at the shear joint to transmit the load of the structure.

Two possible solutions exist: The first is to use two sets of cables—a principal and an auxiliary set, as shown in Figure 8.7. The principal set is deployed to carry the final loading of the structure; the auxiliary set will be deployed in such a way that when it is loaded in balance with the principal set, it will compress the ring without introducing any loading to other structure members. After the ring is compressed and bonded to the base, the tension in the auxiliary cable will be altered in conjunction with that in the principal cable, thereby transferring the bearing pressure under the base, while the stress of the various parts of the structure is regulated.

The second method involves using the principal set of cables, to-

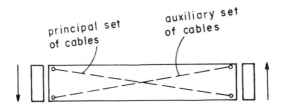

Fig. 8.7. Cable layout of Scheme I.

gether with compressible, coupling packing materials, such as lead, plastic, or steel tubing, to provide a temporary coupling at the gap between ring and base. Narrow strips of packing material i.e., 5 centimeters wide placed 50 centimeters apart, as shown in Figure 8.8 should be adequate to transmit the shear stress when the cables are tightened to transfer the bearing load and compress the ring by allowing the gap to shrink to a predetermined size. The remaining space in the gap may then be grouted permanently with cement. This process is followed by additionally tightening the cable to complete transferring the bearing load. This second method appeared to be less expensive than the first and was therefore recommended.

Optimum Loading of the Tower Base

According to the scheme described above, it is possible to adjust (within certain limits) the radial pressure applied to the tower base in conjunction with the shear loading needed to transfer excess bearing pressure beneath the tower. What then is the proper radial pressure to apply to the tower base to match the shear needed to achieve desired structural safety for the stonework?

Figure 8.9 shows the pressure and shear distribution before and after stabilization. The maximum stress of 90 ton/m² at the lower end of the base before stabilization may be considered the "proof stress" of the material and used as the limit to design the load transferring procedure in order to keep the principal stress in the stonework within this limit. Of particular interest is determining the optimum radial pressure required to match the needed shear of 55 ton/m² at the lower end of the

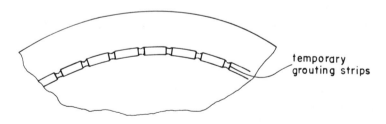

temporary
grouting strips

Fig. 8.8. Gap preparation in Scheme II.

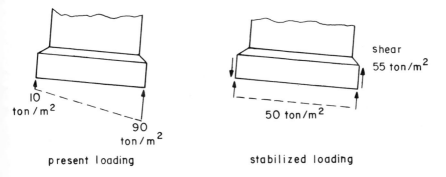

Fig. 8.9. Load distribution before and after restoration.

tower base, when the average pressure is reduced to 50 ton/m² from 90 ton/m².

Figure 8.10 is a two-dimensional Mohr's diagram used to determine the various loading conditions for the tower base with the same fracture capability. Curve *a* represents the proof stress condition with stress along the principle axes at 90 ton/m² and 0 ton/m², respectively.

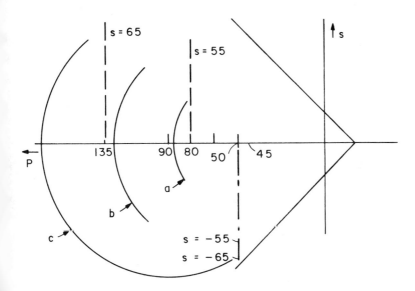

Fig. 8.10. Mohr's diagram used to determine the relationship between pressure and shear loading.

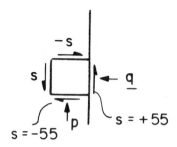

Fig. 8.11. Pressure and shear of an elementary block.

Curve b represents the stress condition of an elemental block (Figure 8.11) near the circumferential surface of the base, shown in Figure 8.6. On the vertical surface of that elemental block are the shear stress of $s = 55$ ton/m² and the matching pressure of q, to be determined; whereas, on the horizontal surface, there are the average compressive stress of 50 ton/m² and a counterbalanced shear $s = -55$ ton/m², as required by the equilibrium condition. Circle b is determined by passing the point representing $p = 50$ and $s = 55$ tangent to the 45 degree line and tangent to circle a. From circle b, the value $q = 80$ is determined from the conjugate point with $s = +55$ ton/m². Curve c represents the determination of $q = 135$ ton/m², when the shear is increased to 65 ton/m², which may represent the maximum value of shear when distributed unevenly over the height of the base to achieve better performance.

Conclusion

From the above analysis, it appears possible to use the proposed concept to stabilize the tower with a fractural stress level at the base comparable to what it is now. This analysis represents a conceptual exposition and detailed layouts of the configuration of the structural elements would have to be modified to accommodate the added feature.

3. CABLE WINDMILL

In the years before reliable nuclear energy can fill the gap left by fast dwindling fossil fuel, solar related energy deserves consideration. C particular interest is wind power which, in many respects, behaves like hydro-power—a form which we know very well how to harness. One major difference between these two forms of power is that in the hydro-power system, a reservoir (which integrates the water power and builds up the head) is placed before the generator, thereby evening out both power supply and loading of the equipment. For the windmill, a reservoir can help even out power demand but cannot help even out loading of the generator. Regardless, a reservoir is needed, if a wind power generating system for public service is ever to be contemplated.

The uneven loading (inherent in the nature of the wind) puts great strain on the structural design. The relatively low power density of wind power, compared to that of hydro-power, implies a general increase in the physical size of equipment. For this reason, we see experimental windmills becoming larger than 100 feet in diameter; imagine a forest of these windmills nodding on the hilltop to supply power needed for a small town.

It is well known that for each structural configuration, there is an optimum range for its adaptation. For bridge design, the configuration of an I-beam is good for a span up to 150 feet, a truss would be comfortable up to 1000 feet, and a suspension bridge is a must for distances beyond 4000 feet. The ordinary configuration for a windmill, with its cantilevered blade, has an upper limit which has probably already been exceeded in current experimental attempts. A cable windmill stretched across a valley where prevailing wind maintains a relatively steady power supply may offer an answer to utilizing wind power for public service, because of structural efficiency comparable to that of a suspension bridge or cable car.

Figure 8.12 shows the possible construction of such a cable windmill. In essence, a set of cables stretched over two giant wheels is suspended across a valley where prevailing winds often exist. A series of airfoil blades are attached to the cable so that the aerodynamic force generated by the blades will move the cable and the wheels which support the cable. A suitable means of generating power may be attached to the wheel shafts to absorb wind power. The advantages of this arrangement for generating power, and especially on a large scale, are:

1. the valley provides a natural venturi to guide and intensify the wind;
2. the sides of the mountain provide the rigid supports needed for the configuration;

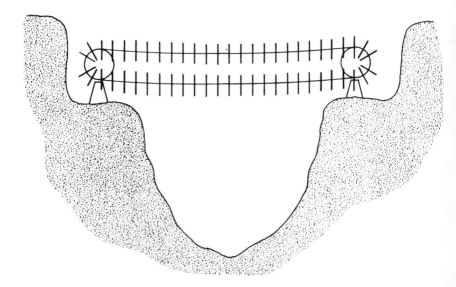

Fig. 8.12. Layout of a cable windmill across a valley.

3. two sets of wheels provide a simple mechanism to support the multiple sets of airfoils with low mechanical energy loss;
4. the terrain provides a natural setting for installing reservoirs at different elevations to be used for storing energy, a needed accessory in large scale, wind power generation.

Detail Design Considerations

Figure 8.13 shows the configuration of the airfoil blades and cables near one end of the cable structure: Two cables give an appropriate amount of torsional rigidity to the suspension system; straddling the cables are a sequence of cross members carrying the posts of the airfoil blades. The blades are arranged symmetrically up and down with respect to the cross members. A number of possible construction methods may be used to fashion the lightweight airfoil blade needed for intercepting the wind to generate power.

The path of the cables and the orientation of the blades reverse as they pass over the wheels; for this reason, it is not necessary to change the pitch angle during steady operation. However, the pitch angle settings of the blades, with respect to the cross members, are changeable

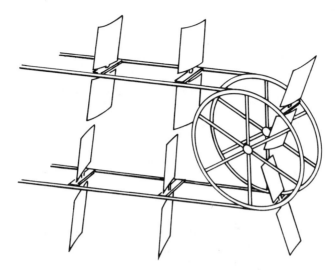

Fig. 8.13. Main drive wheels of cable windmill.

In Figure 8.14, the wheels are mounted to a yoke which is counter-balanced by a set of weights to maintain constant cable tension, a common arrangement in cable chair lift design.

Figure 8.15 shows the top view of the cable suspension system, with provision for supporting the lateral wind loading. In this configuration, a number of wheels are mounted on an equalizing carriage to engage with the cable from the side. The reaction force represented by the spring is in direct proportion to deflection. In Figure 8.15, only one side of the cable is backed by the wheels to resist the wind loading, as represented by the arrow; a similar arrangement may be introduced on the lower side of the Figure.

Estimated Power

Power generated can be estimated as follows:

span	5,000 ft
airfoil spacing	10 ft
number of airfoils	1,000
area per airfoil	1,000 ft^2
nominal power per airfoil	8 kW at 14 mph wind

total nominal power   8,000 kW

weight/per ft^2 2 lb

total weight of blades   2,000,000 lb

slope of cable at wheels   1/4

cable tension $= \dfrac{2,000,000 \times 4}{8}$ lb

size of cable   = 3 in. diameter

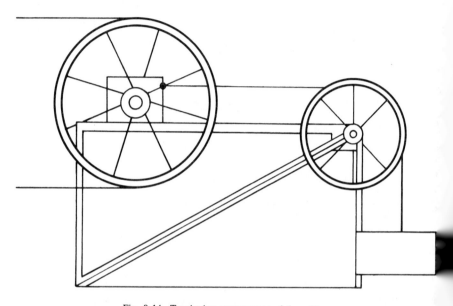

Fig. 8.14. Tensioning arrangement of the cable.

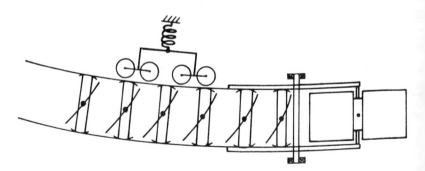

Fig. 8.15. Lateral load compensation.

Pitch Angle Consideration

The effective "angle of attack" of the blade with respect to relative wind velocity is a function of the curvature and speed of the cable and the position of the blade, as shown in Figure 8.16. Assuming the pitch angle β is fixed, then the "angle of attack"

$$\alpha = \theta - \beta,$$

where θ is the angle between the wind and the cable which is largest and most effective when the wind complements the cable movement (left hand side); conversely, the smallest of these two parameters occurs at the other end of the cable, if the pitch angle β is fixed over the entire cable length. It should be adjusted to maximize power output with the idea of varying the angle of attack.

The pitch angle of the blade can be adjusted when the blades are circling around the wheel. The mechanism needed to accomplish this objective is quite similar to that used to maneuver the gondolas at the end of the lift and is well known to those skillful in the art of machine design.

A more elaborate system may involve pitch angle adjustment along the entire cable length, with feedback control and computer programming.

Separation Between Cables

The distance separating the upper and lower cable is controlled by adjusting the differential torque on the two wheels. For instance, in Figure 8.12, if the wheel on the left-hand side is subjected to a higher drive torque than that on the right-hand side, then the upper cable will be stretched tighter than the lower cable, and the two cables will be more widely separated to avoid collision of the blades.

Damping

Failure of the Tacoma suspension bridge, caused by aeroelastic oscillation coupled with strong wind, poses the question whether the cable windmill would face a similar ill fate. The answer, however, is no. While the bridge was a "vibrating reed" type of structure in the form of a semirigid, main deck and susceptible to aero-elastic oscillation and failure, the cable windmill does not have an equivalent "vibrating

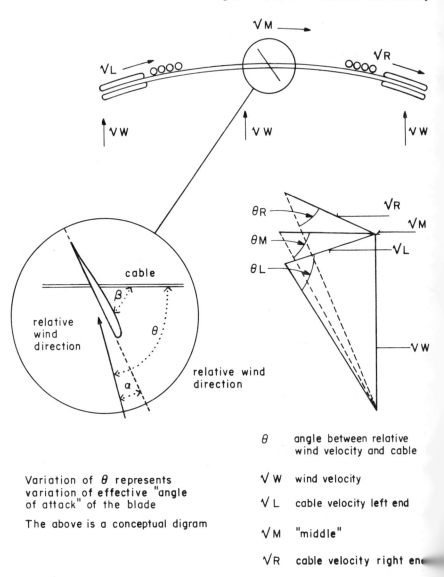

θ   angle between relative
     wind velocity and cable

Variation of θ represents $\sqrt{}$ W   wind velocity
variation of effective "angle
of attack" of the blade $\sqrt{}$ L   cable velocity left end

The above is a conceptual digram
                                 $\sqrt{}$ M   "middle"

                                 $\sqrt{}$ R   cable velocity right end

Fig. 8.16. Analysis of wind loading of cable windmill.

reed'' structure, and the airfoil blades are effective damping elements for the cable.

4. AN INNOVATIVE BOW

For thousands of years, man has tried to improve the characteristics of the bow, which was an ingenious innovation and served him well as basic weapon and sporting device. The basic mechanism is simple and satisfies a fundamental law for converting stored mechanical energy into a high speed projectile efficiently. Efficiency is derived from using the string attached to the tips of the limbs of the bow to transmit stored energy in the limb to the arrow. The parameter worth noting is that, due to the geometrical configuration, the motion of the string is greatly amplified while remaining light in weight, so that kinetic energy loss in the string is quite low.

The same configuration, however, has one drawback, since the force required to pull the bow increases sharply when approaching full draw, a well known fact. This characteristic matches the capabilities of the human anatomy fairly well, because, at full draw, the elbow of the archer's drawing arm folds back 180 degrees, while the other arm extends straight forward. In this position, both arms are in a "dead center" position, giving maximum leverage; even so, it is strenuous to hold a bow in this position while patiently taking aim. Man tirelessly worked for generations to modify the shape of the bow to improve upon this draw force characteristic; the recurve bow probably came closest to being successful, until the emergence of the compound bow a few years ago. One such bow was given to the innovation center by the AMF research center. It served very well as a puzzle for the uninitiated students and faculty who sought to analyze the function, principle, and parameters of that rather complicated looking device. By any standard, it was a first-rate invention, and the inventor made out very well indeed. (The reader may find it in many sporting goods stores and is invited to analyze the principle himself.) Our challenge at the time was to try alternative schemes. Several students attempted various leverage arrangements, most of which turned out to be prior arts.

In the author's "bag of tricks," there was one "nonlinear beam" concept which was quite appealing. This nonlinear beam is a very common device: The pocket ruler employs the principle; i.e., the ruler is quite stiff when extended, due to its built-in camber, yet becomes limp once folded. (The deployable antenna of a space vehicle uses the same principle.) However, it took the author quite a while to realize that the cambered ruler is a bi-stable-state device, since it snaps straight out or bends sharply but will not maintain a position in between.

The solution came about subconsciously, while lecturing in class. The logic established was that if two curved pieces are joined at the inside curvature, as shown in Figure 8.17, the outside curvature opens up in relation to the bend of the curved beam. The cambered ruler described earlier represents a special case of this general configuration when the curvature of the sheet becomes infinity.

This classroom exercise took place the day before Dr. Tom Butler, Director of Research of the AMF Corporation visited the M.I.T. innovation center. That evening, at home, the author thought it would be fun to build a small model of a bow to show to Dr. Butler. The only material he could find was an old celluloid drafting triangle from which two crescent shaped pieces were cut and hinged together with nylon thread and epoxy (Figure 8.18). The outside edges were cut into notches to accommodate a weave of rubber bands. After a string was attached, the model bow clearly exhibited the desired property, with reduced drawing force at full draw.

Dr. Butler was delighted with the new toy that he took back to AMF to show his colleagues. This created some excitement at the Wing Archery Division, because the Olympic games for the following year were supposed to be held in Montreal, an ideal arena for trying out the bow. Furthermore, the compound bow was disqualified from international target competition by a ruling which specified that the string must be attached to the limb directly instead of to a cam. If no new rule were forthcoming, the new bow, with its desirable characteristics and clean appearance (the rubber band used in the model would be replaced by a rubber sheet) might have an excellent opportunity for a debut.

A licensing arrangement was signed by AMF and M.I.T. after long negotiations, which was then followed by an accidental fire in the Wing Archery division, along with the transfer of the manager of that division to another area. Ultimately, the Olympic games ended, and interest in further developing the bow began to taper off.

An analysis of the characteristics of this type of bow is given at the

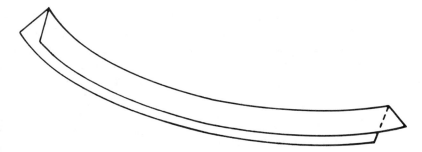

Fig. 8.17. Basic nonlinear beam configuration.

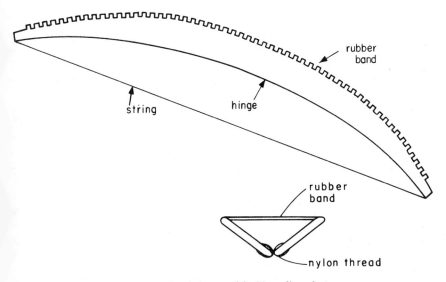

Fig. 8.18. Simple bow model with nonlinear beam.

end of this section. A 14-inch model with 10 pounds of full draw (which dropped down to 5 pounds) was tested by the author. One important side effect of such nonlinear devices is stability. The two halves of the limb must have sufficient torsional rigidity to maintain stability so that the entire limb will open uniformly. The torsional rigidity must nevertheless be achieved without too much rigidity in the lateral bending mode in order to rapidly reduce the drawing force. In a limb constructed of flat sheet, the torsional rigidity increases with the third power of the width, while the bending rigidity increases directly proportional to the width; thus, it is possible to acquire necessary rigidity if sufficient width is present. Aesthetically, however, it may be desirable to use a narrower limb than that required for torsional stiffness, in which event a hollow corrugated sheet may be the answer.

Analytical Design of the "Li" Bow:

1. The Relationship of Displacements at Different Points of the Bow as a Function of Bending

   In visualizing this problem, the designer can benefit greatly by using a piece of cardboard cut in the shape of the surface of a Chinese folding fan; in constructing the bow, the limbs are made of a pair of crescent shaped sheets. Tapering of the tip is due primarily to structural efficiency as a beam.

   The inside curvature of the limb represents the basic mechanism of the three-dimensional, angular coupling of the members.

   Figure 8.19 illustrates in perspective how a designer would visualize the "fan sector" inclining with its inside edge l = length of this edge as indicated) resting on the surface of a table. The fan sector would change its inclined angle θ as the inclusion angle α is changed. Their relationship is governed by the fact that any specified line on the fan surface can bend without changing length; thus, the length of the outer curve l' remains unchanged, while the surface is curled. The detailed relationship of the various dimensions of the sector is shown in Figure 8.19.

2. Graphic Computation of the Relationship Between String Deflection and Spring Deflection

   a. The Limb and the String

   Figure 8.20 shows the relationship of the position of all the important elements of the Li bow through its drawing processes divided into seven stations. Curve $0_1 - 0_3$ represents the original inner arc, with radius R_0 centered at 0_2. As the bow bends, the projected curvature increases, while the radius is reduced. (Pro-

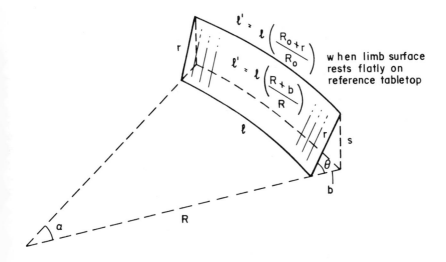

when limb surface
rests flatly on
reference tabletop

$$\theta = \cos^{-1} b/r$$

$$\ell \frac{R_0 + r}{R_0} = \ell \frac{R + b}{R}$$

$$\frac{b}{r} = \frac{R}{R_0}$$

$$\theta = \cos^{-1} \frac{R}{R_0}$$

$$s = \sqrt{r - b} = r\sqrt{1 - \frac{R^2}{R_0^2}}$$

Fig. 8.19. Basic geometry of a section of a fan-shaped beam.

jected curvature means the projection of the curve into the surface of symmetry or the reference plane, as represented by the
tabletop discussed earlier.)

The centers of the curvature (limb arc) shift from 0_2 to C_1, C_2
. . . C_7, along the abscissa $0_1 0_2$. Center C_1 represents the center
of the arc $(0_1 - b_1)$ of the bow, with the string in the initial, un-
drawn position. Chord $b_1 - a_1$ represents the length of one half
of the string.

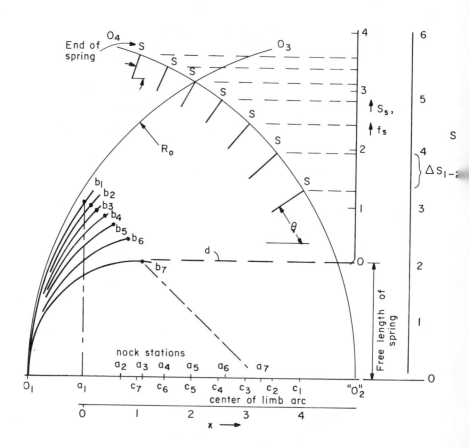

Fig. 8.20. Graphic solution of fan-shaped beam.

In the present analysis, the stations of the draw are laid out with equally spaced centers as reference points. As an illustration, the distance between centers of successive stations is chosen to be half the distance $0_2 - C_1$ of the initial displacement of the center. This distance, as shown, equals $R_0/6$ which corresponds to a wing angle of $\theta = 35$ degrees. These values may be chosen by the designer for his own application.

The broken line $b_1 - a_1$ represents the string in its initial condition; likewise, line $b_7 - a_7$ represents the string at station 7: Similar lines may be shown in between to represent the string at other stations.

The uniform scale shown in the bottom abscissa represents the measurement of deflection of the string at the "nock points" a_1 through a_7, in terms of x.

b. The Wing Angle and the Spring

On the same plot of Figure 8.20 is a second arc $0_2 - 0_4$, to be used to compute the wing angle θ according to the equation given in Figure 8.19:

$$\theta = \cos^{-1} \frac{R}{R_0} .$$

Since the distance $0_1 - C_1$ is R_1, it follows that wing angle θ_1 can be determined by projecting C_1 upward to intersect curve $0_2 - 0_4$ at S_1. The line $0_1 \, S_1$ therefore represents the face of the limb.

Dotted line d represents the "initial" position of the end of the spring. The force f_s generated by the spring is assumed to be proportional to its displacement, which is in turn proportional to the distance S_s measured from the dotted line; thus, a linear scale f_s is provided for this purpose. Paralleling scale f_s, a second scale S is provided, which can be used to measure the movement of the end of the spring, as well as the incremental displacement ΔS.

c. Graphic Computation

Table 8.1 shows the compliance parameters at the spring and the string. By following the law of conservation of energy, we have:

$$\Delta X f_x = \Delta S f_s$$

$$f_x = \Delta S f_s / \Delta X.$$

The result of this table is plotted in Figure 8.21, which shows a peak of f_x at $X = 1$.

This initial try illustrates the various Possibilities for designing the bow. For example, in a bow with inherent stiffness, represented by the dotted line, the final result would be a double bend curve of the broken line, which apparently represents an ideal set of characteristics, because the valley tends to give the archer a sense of controlling the total energy.

Table 8.1.

station	X	f_s	S	F_x	X
1-2	.7	1.6	.62	1.415	.35
2-3	.4	2.2	.52	2.86	.9
3-4	.4	2.65	.40	2.65	1.3
4-5	.5	3.0	.32	1.92	1.75
5-6	.6	3.3	.25	1.39	2.25
6-7	.65	3.52	.2	1.08	2.95

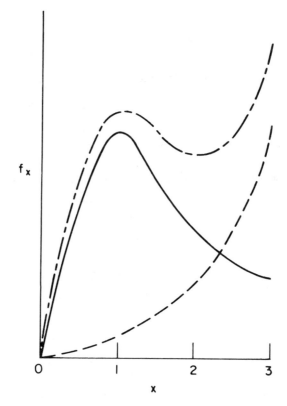

Fig. 8.21. Characteristics of the "Li Bow".

d. Analytical Solution

The analytical solution to the previous discussion follows:

$$L = R\alpha$$
$$l = R_1 \sin\alpha_1$$
$$R = \frac{R_1 \alpha_1}{\alpha}$$

$$X = R(1 - \cos\alpha) = \sqrt{L^2 - (R\sin\alpha)^2}$$
$$f_s = K(\sqrt{R_0^2 - R^2} - S_0)$$
$$S = \sqrt{R_0^2 - R^2}$$
$$f_x = f_x \frac{ds}{dx} = f_s \frac{ds}{dR} \Big/ \frac{dx}{dR}$$

where:

L = half limb length
l = half string length
R = radius of limb
α = half arc angle of limb
R_1 = radius of limb at start
R_0 = radius of curvature of flat limb element
K = spring constant
S_0 = initial spring length
S = spring end movement
f_x = drawing force
f_s = spring force
$\frac{ds}{dR^1}$, $\frac{ds}{dR}$ to be solved from given equations

9
THERMODYNAMICS INNOVATION

Ernest G. Cravalho

1. THERMODYNAMICS OF INNOVATION

As we enter an era of energy constraints, more and more attention must be given to energy considerations in the innovation process. Virtually every innovation consumes energy at some point during its manufacture or its application, and innovations which facilitate energy conversion or conservation will become more prominent in this era. Since the laws of thermodynamics set the ultimate limits on the ways in which we can utilize the energy available in nature, it follows that the science of thermodynamics will play an ever increasingly important role in the innovation process, particularly in energy conversion. Thermodynamics lends itself readily to parameter analysis, and the results derived from this treatment add substantially to our ability to innovate during this challenging period. In this chapter we consider some examples of parameter analysis, and we develop new elements that can be added to our "bag of tricks."

From the point of view of the inventor, the most important feature of thermodynamics is the coupling of the thermal and mechanical aspects of all engineering systems. These engineering systems possess certain characteristics that can best be termed mechanical, in that they are most easily described by properties such as, force, pressure, volume, area, electric charge, etc. These same systems also possess characteristics that can best be termed thermal and described in terms of properties such as temperature and entropy. Not only can the properties of an engineering system be classified as either thermal or mechanical, but we can also describe the interactions that occur between a system and its environment as being either mechanical or thermal in nature. Specifically, mechanical interactions between the system and its environment are termed work interactions, whereas thermal interactions between the system and its environment are known as heat interactions. Obviously these statements need to be refined further, but they are sufficient for the present purposes.

In the context of engineering design or invention, we are concerned with the way in which the thermal and mechanical aspects of system behavior are related. In particular, we are concerned with how the mechanical properties of a system are influenced by thermal interactions, i.e., heat transfer, and how thermal properties of the system are influenced by mechanical interactions, i.e., work transfer. If the mechanical properties are changed by a heat transfer interaction and if the thermal properties are changed by work transfer interaction, we say that the system exhibits thermodynamic coupling. Traditionally, the design engineer takes advantage of this thermodynamic coupling to produce useful work by means of heat transfer interactions from fossil fuels undergoing oxidation reactions (combustion). As we shall show, uncoupled systems are incapable of such energy conversion. Up to the present, little attention has been given to understanding the nature of this coupling and the ways in which it is exploited, but the current concern with energy and its production and use has mandated careful attention to this concept. In the present treatment, we shall attempt to evaluate the nature of thermodynamic coupling and describe some of the ways in which this coupling can be exploited for the purposes of invention.

All engineering systems exhibit thermodynamic coupling to some degree; however, in some cases, this coupling is so weak that, for all intents and purposes, it can be neglected; systems of this type are said to be uncoupled. Their thermodynamic behavior can be easily treated by considering the mechanical and thermal aspects separately. For this purpose, we use system elements, where each aspect of system behavior is described by a particular class of element, and the thermodynamic behavior of the system as a whole is simply the collective behavior of the individual system elements. Thus, it is possible to identify mechanical aspects arising from mechanical interactions and thermal aspects arising from thermal interactions, since these various aspects are localized within a single system element. Of course modeling system behavior in this fashion has its limitations and will eventually break down when the thermal and mechanical aspects become coupled; nonetheless, it is useful for modeling the behavior of a large class of thermodynamic systems.

The pure system elements necessary to describe the more common aspects of the thermodynamic behavior of an uncoupled system fall into three general categories: 1) pure, conservative, mechanical system elements; 2) pure, thermal system elements; and 3) pure, dissipative system elements. These system elements are termed "pure" in the sense that each can experience a specific interaction with its environment, either mechanical or thermal; furthermore, this single interaction manifests itself within the system element as a single phenomenon. That is, the interaction experienced by the system element can occur only in a specific way, so that the change that it produces in the state of the system is always of the same form and always predictable. In addition, the state of a pure system element is particularly simple, since it can be described at any instant in terms of a single, independent property. Any other property of the system can be related to this single, independent property through a mathematical expression known as a constitutive relation. Because of the unique way pure system elements change state, the constitutive relation also describes the path (sequence of states) that the system must follow as a consequence of interactions with its environment. In short, each pure system element exhibits a single characteristic manifested by a unique response to its interaction with its environment.

2. PURE, CONSERVATIVE, MECHANICAL SYSTEM ELEMENTS

The pure, conservative, mechanical system element experiences only work transfer interactions. Thus, the first law of thermodynamics gives the change in stored energy resulting from work transfer:

$$-W_{1-2} = (E_2 - E_1).$$

It is important to note that the thermodynamic sign convention for work transfer requires that work transfer from the system, W_{1-2}, decrease stored energy, E, a property of the system. From this equation, it is clear that when a pure, conservative, mechanical system element is restored to its initial state after having experienced a change, the work transfer associated with return to the initial state must be in the opposite direction of the work transfer associated with the original change of state. Furthermore, the magnitudes of the two work transfers must be identical in order for the system to return to its original state. Because of this "recoverability" of work transfer, the pure, conservative, mechanical system element is termed "conservative." In a conservative system, all energy associated with work transfer into the system is stored within the system and can be later extracted from the system undiminished by returning the system to its initial state. Conversely, the system is incapable of energy conversion in the sense of producing work via heat transfer interactions as a heat engine can. The system can deliver energy only in the form and amount of energy delivered to it at a previous point in time. Thus, when a pure conservative mechanical system element executes a cycle, it experiences no net work transfer, that is, $\oint \delta W = 0$. This means that for a pure conservative mechanical system element, the work transfer is a property of the system. It is to be noted at this point that this is not a general characteristic of thermodynamic systems. Because of the coupling that can exist in thermodynamic systems, it is usually the case that the work transfer is not a property of the system. More attention will be given to this point later in this treatment.

A good example of a pure, conservative, mechanical system element is the pure, translational spring. The pure, translational spring is a model traditionally used to describe the mechanical characteristics of a coil spring where dissipation can be neglected; specifically, a pure, translational spring is a one-dimensional spring of zero mass which can experience only translational displacement along its axis. For such a system element, two significant properties are $\bar{x}$, the relative displacement of the element boundaries at the ends of the spring, and F, the axial boundary force exerted on each end of the spring. Since the pure, translational spring has no mass, the axial boundary force has the same magnitude but opposite direction at the two ends of the spring. These two properties are related by the constitutive relation:

$$F = f(x),$$

where x is now measured relative to the unstressed, relative displacement $\bar{x}_0$ for which F vanishes. From this relation, it is clear that only one of the two properties F and x can be regarded as independent; the choice of which one is independent, i.e., externally imposed, depends on the circumstances of the physical situation.

For an ideal, pure, translational spring, the constitutive relation between the force and the relative displacement is a linear one; thus:

$$F = kx.$$

It follows then that when the pure, translational spring interacts with its environment, a work transfer occurs, and the energy of both system and environment change. In the case of the ideal, pure, translational spring, the constitutive relation gives the path of all possible work transfer processes, such that when it is substituted into the first law of thermodynamics for the spring, the following expression results:

$$-W_{1-2} = \int_{x_1}^{x_2} kx\, dx = \frac{1}{2} kx_2{}^2 - \frac{1}{2} kx_1{}^2.$$

When this expression is substituted into the first law for the spring, the change in stored energy for a pure, translational spring depends upon only the initial and final states of the system. In the ideal case, the change in stored energy becomes:

$$(E_2 - E_1)_{\text{mech}} = \frac{1}{2} kx_2{}^2 - \frac{1}{2} kx_1{}^2,$$

where the subscript "mech" denotes the mechanical nature of the stored energy.

3. PURE, THERMAL SYSTEM ELEMENTS

Suppose now that we desire to represent the thermal characteristics of a coil spring. This may be necessary in order to determine the energy required during the heat treatment process of a coil spring. The pure, conservative, mechanical system element used in the preceding section to describe the mechanical characteristics of the coil spring is wholly inadequate for the present purpose. Note that the pure, conservative, mechanical system element used to describe the mechanical characteristics of the coil spring is capable only of work transfer interactions, but, in the present case, we must evaluate heat transfer interactions. Therefore it is necessary to develop another class of system elements in order to describe the way in which a coil spring behaves during thermal interactions; the pure, thermal system element has been developed specifically for this purpose. A pure, thermal system element is a system element with a single, independent property and has heat transfer as the only mode of interacting with the environment. The two significant properties of a pure, thermal system element are the temperature, T, and the stored thermal energy, U; these two properties are related by a constitutive relation:

$$U = f(T).$$

The first law of thermodynamics may be used to relate the energy change of a pure, thermal system element to the heat transfer interaction at the boundary of the system. Since the pure, thermal system element cannot experience work transfer, the first law of thermodynamics becomes:

$$Q_{1-2} = U_2 - U_1,$$

for a process which takes the pure, thermal system element from state 1 to state 2. For an infinitesimal change of state, the preceding equation can be written:

$$\delta Q = dU,$$

where the symbol δ distinguishes between an infinitesimal interaction and an infinitesimal change of state, as denoted by the symbol, d. For an infinitesimal change in temperature, we can define the heat capacity, C, of a pure, thermal system element as:

$$C = \frac{dU}{dT}.$$

That is, the heat capacity represents the change in the stored, thermal energy per unit change in temperature. We can now define heat capacity per unit mass of the pure, thermal system element as specific heat and denote it by the symbol, c, where:

$$c = \frac{C}{m} = \frac{1}{m}\frac{dU}{dT} = \frac{du}{dT}$$

and u is the stored, thermal energy per unit mass of the system element. In general, specific heat and hence heat capacity are functions of temperature; however, there are many physical situations where specific heat of the pure, thermal system element is essentially constant. Under these circumstances, the stored, thermal energy of a pure, thermal system element becomes:

$$U_2 - U_1 = m(u_2 - u_1) = mc(T_2 - T_1) = C(T_2 - T_1).$$

Then, with the aid of the first law of thermodynamics, the heat transfer experienced by the pure, thermal system element can be expressed by:

$$Q_{1-2} = U_2 - U_1 = mc(T_2 - T_1) = C(T_2 - T_1)$$

Thus, for the pure, thermal system element, heat transfer becomes a property of the system, since the heat transfer required to change the temperature of the pure, thermal system element from the value T_2 back to the original value T_1 is exactly equal in magnitude and opposite in sign to the heat transfer required to change the temperature from the value T_1 to the value T_2. Suffice it to say that in the general case, heat transfer for a thermodynamic system is not a property of the system.

In spite of what seems to be a rather severe limitation of this model, it can be used rather effectively to determine heat transfer required to change the temperature of a real coil spring from one value to another during a heat treatment process.

4. UNCOUPLED THERMODYNAMIC SYSTEM

Since we now have two pure system elements, namely, the pure, conservative, mechanical system element and the pure, thermal system element, it is a simple matter to construct a thermodynamic system which will have both mechanical aspects, characterized by the pure, conservative, mechanical system element, and thermal aspects, characterized by the pure, thermal system element. That is, we can simply take one element of each type and enclose the two together within a system boundary. We then have a system capable of mechanical interactions with its environment, via work transfer with the pure, conservative, mechanical system element and thermal interactions, via heat transfer interactions between the pure thermal system element and its environment. Note however that such a system is uncoupled since the net work transfer and the net heat transfer for any cycle of the system must be equal to zero. That is, the net work transfer for any cycle of the system must be equal to zero as must be the case for the net heat transfer interaction for any cycle of the system. While such a system is useful in certain situations, it is very uninteresting from a thermodynamic point of view since it is impossible for the system to experience a net positive heat transfer for any cycle and as a result produce a net positive work transfer. This is the essence of the lack of coupling in this system.

The uncoupled thermodynamic system does exhibit a feature which, at first sight, might appear to be thermodynamic coupling; however, close inspection reveals that this is not the case. This situation arises when the system experiences a net, negative work transfer. In order to satisfy the first law of thermodynamics, there must be a net, negative heat transfer of equal magnitude. This situation is the result of dissipation; that is, it is possible to dissipate work within the system and to have it appear as a heat transfer. The dissipative characteristics of a system of this type are most easily characterized by a system element described in the following section.

5. PURE, DISSIPATIVE SYSTEM ELEMENT

Pure system elements of this type possess two particular characteristics. First, the net work transfer for these system elements is always negative. The pure dissipative system element is incapable of experiencing a net positive work transfer with another system; work transfer is always *into* a pure dissipative system element. Second, the detailed manner in which the pure dissipative system element interacts with its environment is very important. That is, the kinetics of the work transfer process have a significant influence on the behavior of the system so that the interaction cannot be described in terms of the values of the system properties before and after the interaction occurred. As a class, pure dissipative system elements are incapable of energy storage. In effect they behave as transducers converting work transfer into heat transfer, but they cannot convert heat transfer into work transfer.

A simple example of a pure, dissipative system element is the pure, translational damper. This is a one-dimensional system element of zero mass which can experience only translational displacements along its axis. The two significant properties of the damper are V, the finite velocity of one end of the damper relative to the other end, and F, the axial boundary force exerted on each end of the damper; since the damper has no mass, the magnitude of the boundary force must be the same at both ends of the damper. Thus, the constitutive relation for a pure, translational damper can be written:

$$F = f(V).$$

It is a characteristic of the pure, translational damper that the boundary force at one end of the boundary acts in the same direction as the velocity at that end of the boundary relative to the other end of the boundary; therefore, boundary force, F, and relative velocity, V, always have the same sign. When the relative velocity changes sign, the boundary force reverses direction. Because of this relationship between the boundary force and the relative velocity, work transfer for the pure, translational damper is always negative. This most important point can be seen more clearly, perhaps, in the mathematical definition of work transfer. Since the pure, translational damper is one-dimensional, work transfer can be written as:

$$\delta W = -F dx,$$

but $dx = V dt$. Thus:

$$\delta W = -FV dt.$$

Since F and V always have the same sign and dt is always positive, the preceding equation shows that work transfer for a pure, translational damper is always negative. Thus, work transfer associated with the displacement of the system element boundary is always into the system element.

The equations describing work transfer for the pure, translational damper reveal another significant feature. The explicit calculation of work transfer requires the time history of the state of the pure, translational damper. Consequently, work transfer cannot be represented solely in terms of system properties before and after work transfer interaction. We conclude, then, that work transfer for the pure, translational damper is not a property, as in the case of a pure, conservative, mechanical system element.

As a special case of the pure, translation damper, consider the linear damper where the constitutive relation assumes the form:

$$F = bV,$$

where b is the constant of proportionality, known as the damping coefficient. If this relation is now substituted into the expression for work transfer, we obtain:

$$-W_{1-2} = \int_{t_1}^{t_2} bV^2 dt.$$

Evaluating this integral requires more detailed information than simply the end states of the work transfer process. We must know the state of the system element at every instant of time during the interaction before we can determine work transfer. In distinct contrast to the pure, conservative, mechanical system element, it makes a considerable difference whether the work transfer process for a pure, translational damper was carried out slowly or rapidly.

We now have at our disposal sufficient pure system elements to model virtually all characteristics of a coil spring, for example. In the case of a real coil spring, there is some dissipation; this is sometimes referred to as hysteresis. The coil spring experiences positive work transfer with the environment, but not all of the energy associated with that interaction is stored within the spring in the form of stored, mechanical energy. In fact, a small portion of it is dissipated through a complex, internal process within the material of the spring itself and is converted into heat transfer with the environment.

This system can be modeled as a pure, translational spring, a pure, translational damper, and a pure, thermal system element with constant heat capacity. When this real system experiences a change of state involving a rapid change in x, the process in the model is as follows: Work is transferred across the boundary; part of the work is transferred to the pure, translational spring, increasing its stored, elastic energy; another part of the work is transferred to the damper, resulting in heat transfer to the pure, thermal system element. As a result of this heat transfer, the stored, thermal energy U and the temperature T of the pure, thermal system element are increased. The proper physical interpretation of the model is: Work is transferred into the system by virtue of the change in x; as a result, the system's stored, elastic energy is changed by:

$$E_2 - E_1 = \frac{k}{2} (x_2{}^2 - x_1{}^2).$$

Due to the dissipative nature of the process occurring within the system, the thermal energy U was increased by:

$$U_2 - U_1 = \left\{ \int_{x_1}^{x_2} F dx \right\} - \frac{k}{2}(x_2{}^2 - x_1{}^2).$$

As a result of the increase in U, the temperature of the system increased:

$$T_2 - T_1 = \frac{1}{C} (U_2 - U_1).$$

Although this system has been satisfactorily modeled as an assembly of pure system elements, we must be careful not to attribute physical significance to the internal properties and interactions of the model. These quantities cannot be identified or measured in the real, physical situation.

As an exercise, the reader can show the effectiveness of this approach in understanding and explaining the thermodynamics of the classical Joule experiment, where water enclosed in an insulated container is stirred by a paddle wheel. In this case, energy transferred to the water via work transfer through the motion of the paddle wheel manifests itself as increased temperature in the water and the container-paddle wheel system.

In summary, for the uncoupled system in the absence of dissipation:

$$\oint \delta Q = 0 \; and \; \oint \delta W = 0,$$

so that the first law of thermodynamics is automatically satisfied

$$\oint \delta Q - \oint \delta W = 0.$$

We now show that for a coupled thermodynamic system:

$$\oint \delta Q \neq 0 \text{ and } \oint \delta W \neq 0,$$

but the first law of thermodynamics is still satisfied:

$$\oint \delta Q - \oint \delta W = 0.$$

6. COUPLED THERMODYNAMIC SYSTEMS

To demonstrate the nature of thermodynamic coupling, we consider two limiting cases: (1) a weakly coupled system, a liquid and (2) a strongly coupled system, a gas. The extent of the coupling can be shown by calculating various moduli, two of which are of interest here. The first of these measures (in the absence of dissipation) the fractional change in a thermal aspect, the temperature T, as a consequence of mechanical interaction in the form of a work transfer due to a fraction change in volume v. There is no thermal interaction, i.e., heat transfer during the process. Since the process is dissipationless, it must be reversible. Since it also occurs in the absence of heat transfer, it follows that the entropy, s, of the system must also be constant. Thus, this modulus can be written:

$$\left(\frac{\partial T/T}{\partial v/v} \right)_s = \frac{v}{T} \left(\frac{\partial T}{\partial v} \right)_s.$$

The second modulus is a measure of the fractional change in a mechanical aspect, pressure, as a consequence of a reversible thermal interaction, heat transfer associated with a fractional change in entropy, s, in the absence of work transfer. Thus, the modulus of interest is

$$\frac{s}{P} \left(\frac{\partial P}{\partial s} \right)_v.$$

According to Maxwell's relations,*

$$\left(\frac{\partial P}{\partial s} \right)_v = - \left(\frac{\partial T}{\partial v} \right)_s.$$

* E.G. Cravalho and J.L. Smith, Jr., *Thermodynamics: An Introduction*, Holt, Rinehart, and Winston, N.Y., 1971, p. 10-9.

For the first modulus, we have:

$$\left(\frac{\partial T}{\partial v}\right)_s = -\left(\frac{\partial s}{\partial v}\right)_T / \left(\frac{\partial s}{\partial T}\right)_v.$$

But from Maxwell's relations:*

$$\left(\frac{\partial s}{\partial v}\right)_T = \left(\frac{\partial P}{\partial T}\right)_v,$$

and from the definition of specific heat at constant volume, c_v:

$$c_v \equiv T\left(\frac{\partial s}{\partial T}\right)_v.$$

Then:

$$\frac{v}{T}\left(\frac{\partial T}{\partial v}\right)_s = -\frac{v}{c_v}\left(\frac{\partial P}{\partial T}\right)_v.$$

We also have:

$$\left(\frac{\partial P}{\partial T}\right)_v = -\left(\frac{\partial v}{\partial T}\right)_P / \left(\frac{\partial v}{\partial P}\right)_T.$$

But the coefficient of thermal expansion, β, is defined:

$$\beta \equiv \frac{1}{v}\left(\frac{\partial v}{\partial T}\right)_P,$$

and the coefficient of isothermal compressibility, κ, is defined:

$$\kappa \equiv -\frac{1}{v}\left(\frac{\partial v}{\partial P}\right)_T.$$

Then:

$$\left(\frac{\partial P}{\partial T}\right)_v = \frac{\beta}{\kappa},$$

and:

$$\frac{v}{T}\left(\frac{\partial T}{\partial v}\right)_s = -\frac{v\beta}{c_v\kappa} = -\frac{v}{T}\left(\frac{\partial P}{\partial s}\right)_v.$$

* Op. cit.

Consider the case of a liquid, water at a pressure of 1 atm and a temperature of 77 °F. Then:

$$\beta = 1.1 \times 10^{-4}\ (°F^{-1})*$$
$$\kappa = 4.6 \times 10^{-3}\ (atm^{-1})*$$
$$c_v = 0.998\ (Btu/lbm\ °R)**$$

Then:

$$\left(\frac{\partial T}{\partial v}\right)_s = -\ \frac{(537)(1.1 \times 10^{-4})}{(0.998)\ (4.6 \times 10^{-3})} = -\ 12.87\ \frac{(atm\ lbm\ °R)}{Btu}$$

$$= -\ 35.00\ \frac{°R}{(ft^3\ /lbm)}\ ,$$

and:

$$\frac{v}{T}\left(\frac{\partial T}{\partial v}\right)_s = \frac{-35.00}{(537)\ (62.247)} = -\ 1.05 \times 10^{-3},$$

where:

$$v = \rho^{-1} = (62.247)^{-1}\ (ft^3\ /lbm)***,$$

and the negative sign indicates that the temperature change is in the opposite direction of the volume change. Thus, if the volume were decreased 50% of its initial value (a huge decrease in volume for a liquid), the thermodynamic temperature would increase by 0.05%.

For the case of a gas, consider helium at a pressure of 1 atm and a temperature of 77°F; under these conditions, helium can be modeled with negligible error as an ideal gas with:

$$Pv = RT,$$

where R is the gas constant and:

$$U_2 - U_1 = mc_v\ (T_2 - T_1).$$

Then:

$$\left(\frac{\partial P}{\partial T}\right)_v = \frac{R}{v} = \frac{P}{T},$$

* R.E. Bolz and G.L. Tuve, *Handbook of Tables for Applied Engineering Science*, CRC Press, Cleveland, Ohio, 1976, p. 91
** Ibid., p. 95.
*** Ibid., p. 91.

and:

$$\left(\frac{\partial T}{\partial v}\right)_s = -\frac{T}{c_v}\frac{P}{T} = -\frac{P}{c_v}.$$

Since:

$$c_v = 0.75 \ (\text{Btu/lbm}^\circ\text{R}),^*$$

$$\left(\frac{\partial T}{\partial v}\right)_s = -\frac{1}{(0.75)} = -1.33 \ \left(\frac{\text{atm lbm }^\circ\text{R}}{\text{Btu}}\right)$$

$$= -3.63 \ \left(\frac{^\circ\text{R}}{\text{ft}^3\text{/lbm}}\right).$$

Then,

$$\frac{v}{T}\left(\frac{\partial T}{\partial v}\right)_s = -\frac{Pv}{Tc_v} = -\frac{R}{c_v} = -\frac{c_p - c_v}{c_v} = (1 - \gamma),$$

where the ratio of specific heat, c_p/c_v, is denoted by the symbol γ. For helium:

$$\gamma = 1.66^{**},$$

and:

$$\frac{v}{T}\left(\frac{\partial T}{\partial v}\right)_s = -0.66.$$

Thus, if the volume were decreased 50% of its original value, the thermodynamic temperature would increase 33%.

If we take the ratio of the moduli for these two cases to be a relative measure of the strength of the coupling, we see that gaseous helium is 630 times more strongly coupled than liquid water.

The second modulus shows the liquid and the gas to be more similarly coupled; that is,

$$\left(\frac{\partial P}{\partial s}\right)_v = -\left(\frac{\partial T}{\partial v}\right)_s .$$

Then for water:

$$\frac{s}{P}\left(\frac{\partial P}{\partial s}\right)_v = 12.87 \ (0.93) = 11.95,$$

* R.E. Bolz and G.L. Tuve, *Handbook of Tables for Applied Engineering Science*, CRC Press, Cleveland, Ohio, 1976, p. 45.
** Ibid., p. 45.

where:

$$s = 0.93 \ (\frac{\text{Btu}}{\text{lbm }^\circ\text{R}})*.$$

For the helium,

$$\frac{s}{P} (\frac{\partial P}{\partial s})_v = 1.33 \ (9.23) = 12.31,$$

where:

$$s = 9.23 \ (\frac{\text{Btu}}{\text{lbm }^\circ\text{R}})**.$$

Thus the ratio of the moduli for the two substances is only 1.03.

Some care must be exercised in interpreting the results obtained for these two moduli, since energy interactions for water and helium differ from one another significantly in both cases. There are, however, some generalizations that can be made. Work done on a gas has significant impact on its thermal properties, but work done on a liquid has essentially no effect on its thermal properties. As we will show later, this manifestation of the thermodynamic coupling can be used to significant advantage in the design of energy conversion systems.

Other moduli that quantify the degree of coupling can be defined, but their significance and interpretation will depend on the application. In some instances, strong coupling can be helpful, while in others it can be disadvantageous. For this reason, it is worthwhile to consider some typical applications of interest to the inventor. A major application of thermodynamics in the innovation process is in the development of power production and refrigeration systems. In these applications, the system usually operates in a cycle in order to conserve the working fluid; therefore, some comments on such systems are in order.

7. CYCLIC SYSTEMS

The first law of thermodynamics for a cyclic system reduces to:

$$\oint \delta Q = \oint \delta W.$$

---

* W.C. Reynolds and H.C. Perkins, *Engineering Thermodynamics*, McGraw Hill, N.Y., 1970, p. 547.
** Ibid., p. 549.

Thus, in order to produce net, positive work, there must be net, positive heat transfer. The second law of thermodynamics sets the limits on heat transfer for the cycle.

$$\oint \frac{\delta Q}{T} \le 0.$$

To interpret this result, recall that entropy is a property. Then,

$$\oint dS = 0.$$

Thus, the second law simply states that the flow of entropy out of the cycle must be greater in magnitude than the flow of entropy into the cycle by an amount equal to the entropy generated within the cycle. From the point of view of the inventor, this means that since entropy can be transferred out of the system only by heat transfer, irreversibilities within the device must be kept to a minimum in order to reduce the size and hence capital costs of heat transfer equipment. These irreversibilities can usually be traced to heat transfer through finite temperature difference or to dissipative processes occurring as a consequence of viscosity of the working fluid. The influence of viscosity on system behavior can be minimized by giving careful attention to the fluid dynamics within the device.

For the purposes of this discussion, it is convenient to think of a power cycle as being divided into a high temperature portion, where heat transfer with the environment is positive and a low temperature portion, where heat transfer is negative. Then the first law gives:

$$\oint \delta Q = \int \delta Q_H + \int \delta Q_L = \oint \delta W,$$

and it follows that $\int \delta Q_H$, the positive heat transfer portion, should be as large as possible and that $\int \delta Q_L$, the negative heat transfer portion, should be as small as possible in order to maximize work output.

From the point of view of entropy, the entropy change of the working fluid associated with the entropy that flows into the cycle is given by:

$$\int dS_H \ge \int \frac{\delta Q_H}{T_H},$$

while the entropy change of the working fluid associated with the entropy that flows out of the cycle is given by:

$$\int dS_L \geq \int \frac{\delta Q_H}{T_L}.$$

(Keep in mind that T_H and T_L are not fixed temperatures but usually vary over the heat transfer processes.) If we assume that we have eliminated irreversibilities within the cycle (this is not possible, but it is approachable):

$$\int \frac{\delta Q_L}{T_H} = -\int \frac{\delta Q_L}{T_L}.$$

Since T_L is usually very nearly the ambient temperature and since we want $\int \delta Q_L$ to be as small as possible for a given $\int \delta Q_H$, it follows that T_H should be as large as possible.

Thus, we have shown from the first and second laws of thermodynamics, quite apart from any considerations of thermodynamic coupling that may exist in the working fluid, two important considerations of power system (heat engine) design: 1) minimize irreversibilities within the system; and 2) positive heat transfer should occur at as high a temperature as possible. In addition to these design criteria, there are others that can be deduced from considering the coupling of the working fluid.

Most thermodynamic plants process a fluid with strong thermodynamic coupling through an apparatus where heat transfers (actual, not equivalent to combustion) are separated from work transfer. This design feature insures that only the state of the strongly coupled fluid (such as a gas) is cycled, while the thermal state of the essentially uncoupled parts of the apparatus remain virtually unchanged. Heat transfer to the fluid is effected by moving the fluid to a region of higher wall temperatures rather than by increasing the temperature of the wall confining the gas. If the thermal state of the walls confining a gas is cycled in order to effect heat interactions with the gas, the energy changes for the walls are larger than the energy changes for the gas (for normal materials over the range of states of practical importance). Furthermore, since every energy transfer interaction is irreversible, especially if it is at a practical rate, the thermal cycling of any structural members or confining walls will seriously increase the irreversibilities of any thermodynamic plant.

These are two of the basic reasons why all practical, thermomechanical, energy conversion plants employ flow systems. Although w cannot express these generalities quantitatively, we can demonstrate their significance with a specific example. For the purpose of illustrating the limitations of a non-flow thermodynamic plant, we will consider in some detail an ideal-gas Carnot cycle operating entirely within a single cylinder. In our analysis, we consider two models: first, a model for the plant which ignores the thermal aspects of the cylinder; and second, a model which includes the heat capacity of the cylinder. In the entire discussion, we shall consider only an ideal gas Carnot cycle plant which operates reversibly between a reservoir at T_H and a reservoir at T_L.

Consider the performance of our first model. One useful measure of performance is the thermodynamic energy conversion efficiency given by:

$$\eta \equiv \frac{W_{net}}{Q_{add}}.$$

For the Carnot cycle, this becomes:

$$\eta = 1 - \frac{T_L}{T_H}.$$

Since we are considering only reversible cycles in this discussion, each of the models will have this reversible efficiency. There are two other characteristics which are important considerations for a practical plant design: first, the mean effective pressure, which is a measure of the work per cycle; and second, the net work ratio, NWR, which is an indicator of the influence of irreversibilities on the cycle. Recall from elementary thermodynamics that work transfers for the four processes of the Carnot cycle are:

1. adiabatic expansion, 1-2

$$W_{1-2} = Mc_v (T_H - T_L)$$

2. isothermal compression, 2-3

$$W_{2-3} = MRT_L \ln(V_3/V_2) = - MRT_L \ln(V_1/V_4)$$

3. adiabatic compression, 3-4

$$W_{3-4} = - Mc_v (T_H - T_L)$$

4. isothermal expansion, 4-1

$$W_{4-1} = MRT_H \ln(V_1/V_4)$$

The mean effective pressure is defined as:

$$P_{m.e.} = \frac{W_{net/cycle}}{V_{displaced/cycle}} .$$

Then for the Carnot cycle:

$$P_{m.e.} = \frac{MR(T_H - T_L) \ln(V_1/V_4)}{V_2 - V_4} .$$

Thus, the mean effective (m.e.) pressure varies directly with the mass of the charge in the cylinder. However, as this mass is increased, the maximum pressure of the cycle also increases, which in turn requires increased strength and weight of the mechanism. As a first order approximation, we assume that the mass of the apparatus varies directly with the maximum pressure of the cycle; therefore, a reasonable measure of specific work of the machine, i.e., work per unit mass of machine, is the ratio $P_{m.e.}/P_4$. Substituting the ideal gas constitutive relation into the expression for mean effective pressure, we obtain

$$\frac{P_{m.e.}}{P_4} = \frac{[1 - (T_L/T_H)]}{\left[\dfrac{V_2}{V_1} \dfrac{V_1}{V_4} - 1 \right]} \ln(V_1/V_4).$$

This result shows that for a fixed temperature ratio T_H/T_L, the ratio $P_{m.e.}/P_4$ depends only on the volume ratio during isothermal expansion, V_1/V_4, since V_2/V_1 is fixed by the temperature ratio of the isentropic processes which are utilized to change the temperature in the cycle. As shown in Figure 9.1, the ratio $P_{m.e.}/P_4$ reaches a maximum at $V_1/V_2 = 2.57$. In geometric terms, this value gives a pressure volume diagram for the cycle which has the maximum pressure difference (between expansion and compression) for its length; that is, it is the "fattest" diagram for the cycle. However, even at maximum pressure difference, mean effective pressure is only slightly over 4% of maximum pressure.

This very low utilization of the displaced volume results from the very sharp pressure rise up to state 4, as seen in Figure 9.2.

The net work ratio for the cycle is defined as the ratio of net shaft work transfer to gross positive shaft work transfer. This parameter measures the fraction of gross positive work transfer produced by the cycle which is available for useful purposes. The fraction not available is expended by negative shaft work transfer components of the plant; thus, a low value for the net work ratio means that a large amount of work produced by the plant recirculates within the plant.

For the Carnot cycle, the net work ratio, NWR, is given by:

$$NWR = \frac{R(T_H - T_L)\ \ln(V_1/V_4)}{RT_H\ \ln(V_1/V_4) + c_v(T_H - T_L)}$$

$$NWR = \eta\left[\frac{1}{1 + \dfrac{1}{\gamma - 1}\dfrac{\eta}{\ln(V_1/V_4)}}\right]$$

As shown in Figure 9.1, the net work ratio increases rapidly with the volume ratio V_1/V_4. At the maximum, mean effective pressure, the net work ratio is about 0.34, which is a bit less than half of the asymptotic value 0.725. Thus, some compromise must be reached between the size of the system and the effect of irreversibilities.

Our second model illustrates the most serious, practical difficulty with the closed system Carnot cycle—namely, the gas must be both heated and cooled by heat transfers from the same cylinder walls. Since the gas must have positive Q, when its temperature is high and negative Q, when its temperature is low, the cylinder walls must be made to change in temperature over the same range as the gas. As we shall see, heat transfers required to change the temperature of the cylinder walls are considerably larger than those required to change the temperature of the gas.

Since the cycle must operate reversibly while in communication with only two heat reservoirs, the system in the cycle must be the gas plus the piston-cylinder apparatus confining the gas. For reversibility, gas and piston-cylinder must be in equilibrium at all times. Including structural parts in the system does not change the pressure-volume-termperature relation:

$$PV = M_gRT,$$

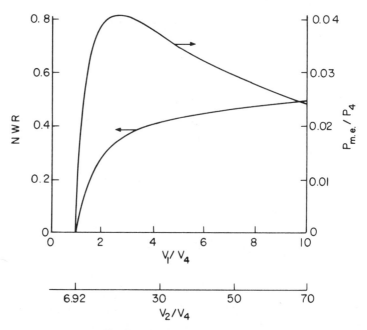

Fig. 9.1. Carnot cycle performance.

where M_g is the mass of the gas in the cylinder. However, the energy of the system must include the thermal energy of the structural parts, which are modeled as pure, thermal system elements and rigid, non-elastic mechanical elements, as well as the internal energy of the gas. Then:

$$U = M_g \ [c_v + \frac{M_s}{M_g} \ c] \ T \cdot$$

where M_s is the mass of the structural parts and c is their specific heat. The system thus behaves as a gas with the gas constant R of the gas alone but an augmented specific heat, $(c_v)_{\text{eff}}$.

$$(c_v)_{\text{eff}} = [c_v + \frac{M_s}{M_g} \ c] \ \cdot$$

The effective ratio of specific heats, γ_{eff}, for the system is:

$$\gamma_{\text{eff}} = \frac{R}{(c_v)_{\text{eff}}} + 1 = \frac{1}{\frac{c_v}{R} + \frac{M_s}{M_g} \frac{c}{R}} + 1,$$

and the results of our analysis for the first model can be used directly.

We can now show that the term $M_s/M_g \cdot c/R$ is a number large relative to c_v/R so that γ_{eff} is only slightly greater than unity. As γ_{eff} approaches unity, the system approaches uncoupled behavior, and NWR and $P_{\text{m.e.}}/P_4$ both approach zero. A simple, first-order estimate of the ratio M_s/M_g can be obtained by considering the cylinder wall as the only contribution to the mass of the structure. The mass of the structure is then limited by the maximum stress σ that the cylinder wall material can withstand. The hoop stress in the cylinder is:

$$\sigma = P \; \frac{D}{2t},$$

where the symbols are defined in Figure 9.3. The volume V_S of structural materials is:

$$V_s \;=\; \pi DtL \;=\; \frac{\pi D^2 LP}{2\sigma}.$$

Thus the mass of the structure is:

$$M_s \;=\; \frac{\pi D^2 LP \rho_s}{2\sigma},$$

where ρ_s is the density of the cylinder material. The mass of gas in the cylinder is from the the ideal gas constitutive relation:

$$M_g \;=\; \frac{\pi D^2}{4} \, L \, \frac{P}{RT},$$

so that the ratio M_s/M_g becomes:

$$\frac{M_s}{M_g} = \frac{2\rho_s RT}{\sigma}.$$

The limiting value of this ratio will occur at the highest temperature where the allowable stress is lowest. As a typical value for this ratio, consider a steel cylinder and helium gas, for which the relevant properties are:

$$\rho_s \quad = 487 \ \text{lbm/ft}^3$$

$$\sigma \quad = 20{,}000 \ \text{lbf/in.}^2$$

$$c \quad = 0.113 \ \text{Btu/lbm} \ °\text{R}$$

$$R \quad = 386 \ \frac{\text{ft lbf}}{\text{lbm} \ °\text{R}} = 0.496 \ \frac{\text{Btu}}{\text{lbm} \ °\text{R}}$$

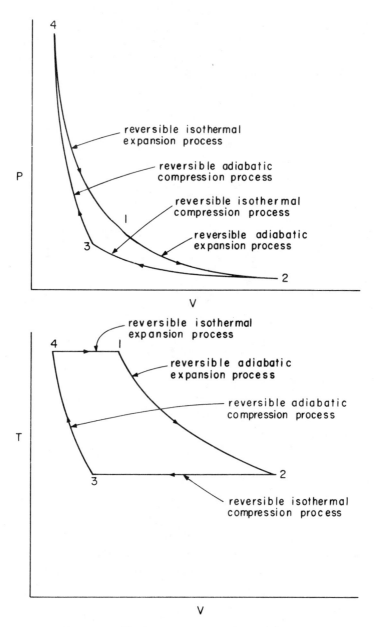

Fig. 9.2. Ideal gas Carnot cycle.

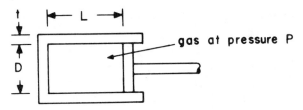

Fig. 9.3. Piston and cylinder.

$$c_v = 1.5\,R$$

For these values and a temperature $T = 2000$ °R, the ratio M_s/M_s has the value:

$$M_s/M_g = \frac{2(487)\,(386)\,(2000)}{(20{,}000)\,(144)} = 261 \cdot$$

The effective specific heat ratio is then:

$$\gamma_{\text{eff}} = 1 + \frac{1}{1.50 + 59.0} = 1.0165 \cdot$$

From the expression for the path of an isentropic process ($PV^\gamma = $ constant), we find that with this very low specific heat ratio, a volume ratio V_2/V_1 of about 2300 is required for $T_H = 2000$ °R and $T_L = 550$ °R. Then the ratio $P_{\text{m.e.}}/P_4$ is reduced to about 10^{-4} at $V_1/V_4 = 2.72$ and the NWR at this same value of V_1/V_4 is reduced to 1.6%. These values show that thermally cycling the mechanical structure of the Carnot engine cylinder makes the closed system Carnot engine completely impractical. Even though theoretically the engine still has reversible efficiency, the small NWR indicates that a very small irreversibility will drastically reduce thermal efficiency.

The two examples presented here show that the impracticability of a thermodynamic plant can be deduced from a parameter analysis based on reversibility without detailed consideration of the rate processes. In the first model we saw that low mean effective pressure for the Carnot cycle renders that cycle completely impractical in spite of its relatively high thermal efficiency. In the second model we saw that cycling of the cylinder walls further reduced the effective thermodynamic coupling of the gas with concomitant reductions in the mean effective pressure and the net work ratio. The implications for innovation in energy conversion are obvious. In particular, the energy conversion efficiency of the cycle (also known as the thermal efficiency) is not the single most important parameter to be considered in evaluating system performance. The mean effective pressure is a measure of the effectiveness of the cycle in using the available hardware, and the net work ratio is a measure of the sensitivity of the cycle to irreversibilities and the size of the system. The lower the value for either of these parameters, the poorer the system performance and the more expensive the capital equipment per unit of energy output.

From the preceding discussion, it is also clear that there are significant advantages to be gained by performing the various interactions in a cycle in dedicated apparatus; that is, the system is composed of a collection of devices, each performing a special function, and the working fluid is simply transported from one device to another. Figure 9.4 shows a Carnot cycle performed in precisely this fashion. The positive heat transfer process, a-b, occurs in one device (high temperature heat exchanger); the positive work transfer process, b-c, occurs in another device (expander); the negative heat transfer process, c-d, occurs in another device (low temperature heat exchanger); and finally, the negative work transfer process, d-a, occurs in still another device (compressor). There are obvious advantages in having the cycle fall within the two-phase region, e.g., the heat exchangers are simple, constant pressure devices rather than complex systems whose temperatures and pressures must be carefully matched throughout the heat transfer processes.

For this system, we can now show some of the important aspects of thermodynamic coupling within the fluid. Consider the expander: Since the operation of the expander in the limiting case is reversible and adiabatic and, hence, isentropic and since the work of the device is given by the change in enthalpy of the fluid as it passes through the device, we would like to have a working fluid for which $(\frac{\partial h}{\partial P})_s$ is as large as possible. Since:

$$dh = T\,ds + dP,$$

it follows that:

$$\left(\frac{\partial h}{\partial P}\right)_s = v.$$

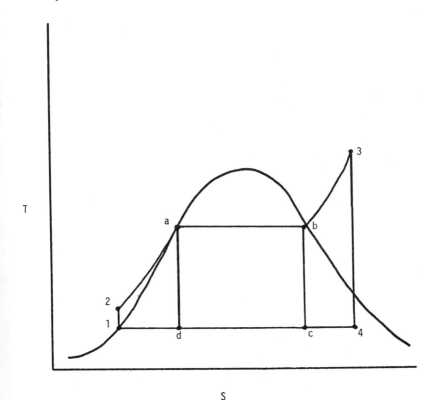

Fig. 9.4. Comparison of Carnot and Rankine cycles.

Thus, we would like to have a working fluid with as large a specific volume as possible. Obviously there are limits, since a large specific volume means a large system and, hence, a large capital investment in plant equipment.

Conversely, for the compressor, we would like to have v as small as possible; operation in the two-phase region precludes this. In fact this is the reason why the network ratio is so small for the Carnot cycle. It was Rankine who observed this phenomenon and suggested shifting the compressor operation into the liquid region, where v is orders of magnitude smaller. The net result was an innovation in the thermodynamic power cycle design, the Rankine cycle, where negative work transfer operations are carried out with a fluid that is weakly coupled, while the positive work transfer operations are carried out with a fluid that is strongly coupled. There is a sacrifice in cycle efficiency, since the average temperature for positive heat transfer is reduced, but the substantial gain in net work ratio more than compensates for this loss. In fact, this loss can be offset somewhat by continuing the positive heat transfer process into the superheat region. The Rankine cycle, complete with superheat, is shown as cycle 1-2-3-4-1 in Figure 9.4. The impact of these recommendations, based on parameter analysis, has been one of the most significant of any innovation in the field of energy conversion.

Consider for example the Carnot cycle a-b-c-d-a of Figure 9.4 with $T_H = 1100°R$, $T_L = 550°R$ and H_2O as the working fluid.

$$\eta = 1 - \frac{T_L}{T_H} = 1 - \frac{550}{1100} = 0.50$$

$$NWR = \frac{(h_b - h_c) + (h_d - h_a)}{h_b - h_c} = 1 + \frac{h_d - h_a}{h_b - h_c}$$

$$= 1 + \frac{474.2 - 679.1}{1133.7 - 701.4} = 0.53$$

For the Rankine cycle 1-2-b-c-1 of Figure 9.4 operating between the same two temperature extremes,

$$\eta = \frac{\oint \delta W}{Q_H} = \frac{(h_b - h_c) + (h_1 - h_2)}{h_b - h_2}$$

$$\eta = \frac{(1133.7 - 701.4) + (58.018 - 64.15)}{1133.7 - 64.15} = 0.40$$

$$NWR = \frac{(h_b - h_c) + (h_1 - h_2)}{h_b - h_c} = 0.99$$

Thus, by shifting the negative work transfer parts of the cycle into a region of working fluid behavior that is essentially uncoupled in the thermodynamic sense, we have nearly doubled the net work ratio with a small sacrifice in energy conversion efficiency. This means that those components of the cycle devoted to negative work transfer can be reduced in size with commensurate savings in capital equipment.

While the sacrifice of energy efficiency can be reduced by superheating (process b-3 in Figure 9.4), there is a limit to the extent to which this heat transfer process can be extended, however. The metallurgical characteristics of typical expanders are such that peak temperatures are limited to approximately 1000°F. Then in order to improve the performance of the system, we must consider operation at higher pressures which will have the effect of increasing the temperatures at which the positive heat transfer occurs. In moving to higher pressures we must keep in mind that the enthalpy of the fluid will decrease as pressure increases at constant temperature. That is,

$$(\frac{\partial h}{\partial P})_T = v - T (\frac{\partial v}{\partial T})_p < 0.$$

The effect that this modulus has on expander output can be determined only after computing a similar modulus for the low temperature heat exchanger.

Coupling can also have a significant effect on losses in the system; that is, in system design, special care should be given to those components most sensitive to system losses. For example, in the case of piping losses, the first law of thermodynamics applied to a fluid flowing through a pipe with no heat or work transfer and negligible changes in kinetic and potential energies shows that enthalpy of the fluid remains constant. As a consequence of thermodynamic coupling in the fluid, pressure drop in the direction of flow generates entropy in the fluid. Thus,

$$(\frac{\partial s}{\partial P})_h = - (\frac{\partial h}{\partial P})_s \Big/ (\frac{\partial h}{\partial s})_P .$$

But since $dh = T\, ds + v\, dP$,

$$(\frac{\partial h}{\partial P})_s = T \text{ and } (\frac{\partial h}{\partial s})_P = v.$$

Then

$$(\frac{\partial s}{\partial P})_h = - \frac{v}{T}.$$

Since v and T are both positive, it follows that the entropy generation must also be positive; it also follows that the entropy increase for a given condition of temperature and pressure drop varies directly with the specific volume of the fluid. Thus, piping design for the transport of low density fluids, such as vapors and gases, requires more attention and innovation than for high density fluids, such as liquids.

Parameter analysis in thermodynamics is not confined soley to energy conversion systems. This approach is valid also for refrigeration systems. The performance of all refrigeration systems, regardless of the means of producing the refrigeration effect, is, like energy conversion systems, dictated by the first and second laws of thermodynamics. Typically a refrigeration system (except for the absorption system which is a special case) operates between two temperatures: the ambient temperature of the surroundings and the temperature of the refrigeration load, i.e., the space or substance being refrigerated. Thus, the only way entropy can flow into or out of the refrigeration cycle is by means of the heat transfers that occur while a unit mass of refrigerant is in thermal communication with the environment and the refrigeration load. The only other phenomenon that contributes to the entropy change of the refrigerant as it negotiates the cycle is the generation of entropy through irreversibilities that occur within the refrigerant. Then for the case of reversible operation as we showed for the energy conversion systems

$$- \int (\frac{\delta Q}{T})_{\text{environment}} = \int (\frac{\delta Q}{T})_{\text{refrigeration}}$$

Since by convention heat transfers into a system are positive and heat transfers out of a system are negative, and since $T_{\text{environment}} > T_{\text{refrigeration}}$, it follows that the heat transfer with the environment must be greater in magnitude than the heat transfer with the refrigeration load. Thus since $T > 0$, the net heat transfer for the cycle must be negative, i.e. out of the refrigerant and into the environment, viz., $\oint \delta Q = \int (\delta Q)_{\text{environment}} + \int (\delta Q)_{\text{refrigeration}} < 0$. It follows then from the first law that the net work transfer for the cycle must be also negative, and since by convention work transfers into a system are considered negative, the cyclic refrigerator can produce a refrigeration effect only by consuming work, i.e., mechanical energy. It also follows from the last two equations that for a given refrigeration load, the lower the refrigeration temperature, the greater the expenditure of mechanical energy. The traditional measure of effectiveness of a refrigerator is the coefficient of performance (COP) which is defined as the refrigeration effect per unit of mechanical energy expended, i.e.,

$$COP = \frac{Q_{\text{refrigeration}}}{-\oint \delta W}$$

where the negative sign is included in the denominator in order to make COP positive. Typical values for COP range from approximately 4 for a household refrigerator to about 0.01 for a refrigeration load at the normal boiling point of helium (4.2 K).

The vast majority of refrigeration systems operate on the vapor compression cycle or some variant thereof. In these systems, heat transfer interactions occur at two different temperatures which by virtue of the thermodynamic relationship between temperature and vapor pressure manifest themselves as two different pressures during the cycle. In order to change pressure of the refrigerant from the low value that exists on the low temperature side of the cycle to the high pressure on the high temperature side of the cycle, a compressor is used, and since the compression occurs while the refrigerant is in the vapor phase, systems of this type are called vapor compression systems.

In the operation of the vapor compression cycle, saturated vapor enters the compressor at low temperature and low pressure and is compressed to high temperature and high pressure. Because the compression process occurs while the refrigerant is in the vapor phase (strongly coupled thermodynamically), the work requirements are quite high and have made vapor compression refrigeration impractical in areas where electricity is not readily available. Fortunately, it is possible to achieve a refrigeration effect without the expenditure of work by employing a system known as an absorption refrigerator. The absorption refrigerator is an innovation that provides an excellent example of parameter analysis.

From the first law of thermodynamics we have already shown that the net heat transfer for a cycle device is equal to the net work transfer.

$$\oint \delta Q = \oint \delta W$$

If the net work transfer is to be zero, the net heat transfer must also vanish. However, we have from the second law of thermodynamics

$$\oint \frac{\delta Q}{T} \leq 0$$

Thus, if we allow heat transfer at only two temperature levels (a heat source and a heat sink, for example), there is no way that we can have $\oint \delta Q = 0$ and still satisfy the second law. However, with the use of three or more levels of temperature, it is possible to have $\oint \delta Q = 0$ and to satisfy the second law. The approach used in the absorption refrigerator is to employ three different temperature levels, a high temperature heat source (fossil fuel flame), an intermediate temperature level (environment), and a low temperature level (refrigeration load). While the concept is simple enough in principle, it does require considerable ingenuity to translate into working hardware.

The key parameter that makes absorption refrigeration cycles possible is the solubility of solutes in solvents and the way in which the solubility is related to temperature and pressure of the solution. This behavior is in fact another manifestation of thermodynamic coupling. A solute which is readily soluble in a solvent at low temperature and low pressure becomes less soluble at high temperature and high pressure. Thus, the vapor phase in equilibrium with the solution at high pressure is proportionately richer in solute. The solute is thus at high pressure and can now be used in refrigeration loop consisting of condenser, throttle valve, and evaporator. A schematic diagram of a basic absorption system is shown in Figure 9.5.

In this cycle, low pressure refrigerant vapor is converted to a liquid phase in solution while still at the low pressure by means of absorption by the absorbent. Because of an affinity between absorbent molecules and refrigerant molecules, the refrigerant dissolves readily in the absorbent. During this process, energy is released by virtue of the heat of condensation, sensible heat, and heat of dilution. This energy is usually rejected by means of heat transfer with the environment. The refrigerant

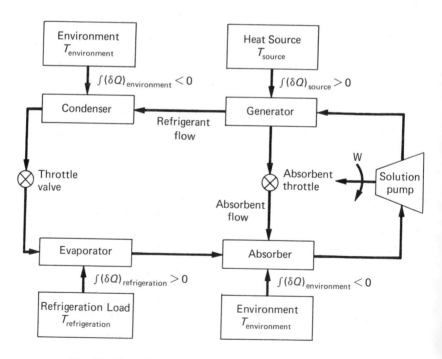

Fig. 9.5. Schematic representation of basic absorption refrigeration cycle.

in the liquid phase is now compressed (along with the absorbent) in the solution pump, and since the refrigerant is in the liquid phase which is very weakly coupled thermodynamically, the work of this compression process is very much less than would be required in a vapor compression cycle with a mechanical compressor. The refrigerant and absorbent in the high pressure solution are then separated in the generator by virtue of a distillation process which requires heat transfer with a high temperature heat source such as a flame produced by combustion of a fossil fuel. In order to enhance vaporization in the evaporator, the refrigerant leaving the generator should be essentially free of absorbent, and the complexity of the hardware required to accomplish this in the generator will be dictated by the relative volatilities of refrigerant and absorbent. However, certain practical considerations such as the need to avoid excessively high temperatures dictate that some refrigerant should leave the generator in the absorbent flow still in solution. As shown in Figure 9.5, the absorbent and refrigerant have different fluid circuits, and the absorbent serves merely as a carrier fluid for the refrigerant stream during the compression process.

The analysis of the performance of an absorption refrigeration cycle can be carried out using the expressions introduced earlier for the mechanical vapor compression systems; however, the heat transfer into the generator from the high temperature source and the heat transfer from the absorber to the environment must be included. The COP of these systems is significantly lower than the COP of a comparable vapor compression system, but when the efficiency of the electrical power generation and transmission network is included in the vapor compression system, the COP's of the two systems are nearly identical.

In this era of energy constraints, the concern for energy conservation has caused considerable interest in absorption refrigeration systems since the sun can be used as the high temperature source via a solar collector. Recent developments in solar collector design have made the absorption refrigeration system competitive with mechanical vapor compression systems.

8. CONCLUSION

From the preceding discussion, it is clear that the laws of thermodynamics and the thermodynamic coupling that exist with working fluids can be exploited for the purposes of innovation. This process can be facilitated by reorganizing thermodynamic knowledge into a form that lends itself to the thought processes of innovation. The approach presented here is but a first attempt at this process. Clearly, more work remains to be done, but hopefully the present treatment has provided a few suggestions to start the innovator in the right direction.

9. COMMENTARY BY Y. T. LI

In this chapter, Cravalho has identified a few key parameters in thermo-dynamics which are closely related to the building block parameters in engine system configuration. The next task should be to examine how these key parameters are actually coupled to some existing system configuration and to even try to "invent" new ones. While this work is being rushed to print, a brief comment on this topic might be of help to the reader. Of those system configurations which might be of particular interest are:

- some new forms of working media, such as Nitinal, a nickel-titanium alloy which exhibits an abrupt, dimensional change over a very narrow temperature gradient;
- solar energy or other forms of low grade heat energy to provide a low cost power supply;
- the Stirling engine, a hybrid of the internal combustion engine with a potential advantage of low exhaust pollution level.

A brief discussion of the first two applications follows.

In discussing the Cyclic Systems, Cravalho addressed the concept of flow systems, emphasizing the desirability of moving the working medium from a heat transfer station to a work transfer station. In so doing, the rigid structure confining the working medium at the heat transfer station need not be subjected to the complete temperature cycle, thereby avoiding a loss due to the irreversible thermal process of the structure. (The temperature cycle of the structure at the work transfer station is unavoidable but is rendered less objectionable through the action of the work transfer station as a thermal insulator.) If Nitinal is adequate as a working medium, assuming that it possesses strong coupling (thermal-mechanical) over the narrow temperature range (an efficient use has yet to be invented), then the flow system concept does not apply, because no container is needed for Nitinal.

For low grade heat, such as solar energy, utilizing the temperature gradient at various ocean depths results in a low build-up of pressure in the working medium and, for this reason, it may be possible to use a "membrane" type container to confine the working medium, i.e., Freon. In this manner, the structure may be sufficiently light so that its thermal capacity is not significant. Furthermore, since, in solar energy, heat energy is transmitted in radiant form, a "transparent" container would admit this energy with low loss, and, accordingly, a solar engine in panel form was conceived. The panel is constructed like an endless belt with a sequence of pockets in which suitable working fluid is entrapped. One side of the belt is made of a transparent plastic sheet to admit the sunbeam. The belt is structured to contract when the working medium expands under the sun's radiation. The endless belt is mounted on a set of rollers, so that when the front is exposed to the sun as a solar panel, the back is cooled by some convenient means, such as a spray of water. A differential pulley arrangement is incorporated into the system so that continuous contraction of the belt on the side facing the sun and relaxation of the belt on the shady side would cause the belt to revolve and deliver a mechanical power output from one of the shafts of rollers supporting the belt.

In the above scheme, the thermal capacity of the thin-walled container may not be a problem, and the transparent container for receiving solar energy also appears to be appropriate. However, a major difficulty may exist in the cooling side: At the cooling side, the working medium is in vapor form and enclosed in separate inflated bags; thus, in each bag, the vapor is cooled primarily by the convection process inside the bag, which is very slow. An uninnovative approach would be to put a fan in each bag to circulate the vapor. Thus in a conventional condensor used for cooling, the function of the flow system identified by Cravalho is not only an important feature for bringing the working fluid to that station, but the flow itself is an important mechanism for promoting heat transfer.

In the above innovative exercise, the goal was to generate a simple solar engine to act as apower supply in isolated (i.e., agricultural) regions. The belt-shaped solar panel seems to be a straightforward configuration satisfying this need, though the cooling problem must be resolved. Like all innovation processes, solutions are spurred by need, processed by matching need with technological configurations, and refined by identifying adverse effects. The response to adverse effects is further innovation, until the final product is completed with all desired features realized. Thus, while the belt-shaped panel is not yet practical for use as a solar engine, it serves to illustrate the innovation process in thermodynamics.

10
EVALUATION, FINANCING, AND DEVELOPMENT OF THE INNOVATION CONCEPT

1. CAPITAL INVESTMENT AND THE RISK FACTOR IN INNOVATION DEVELOPMENT

The development of a new product from an innovative idea invariably requires capital investment at various stages. Investment is usually attracted by potential gain, should the innovation succeed, versus the risk of failure along the development route.

Each stage of new financing invariably involves a dialogue between one who is soliciting and another who is providing the investment. In principle, it is the entrepreneur's or innovator's function to solicit the investment and his responsibility to carry the development to the next stage to realize potential gain. The function of the investor, on the other hand, is to evaluate the proposal made by the entrepreneur by examining the risk factor inherent in development versus the potential gain, in order to assess the intrinsic value of the investment in terms of the equity distribution of the venture.

The information flow diagram in Figure 4.8 of Chapter 4 illustrates the mode of operation of innovation management. Emphasis is placed on developing a conceptual model to guide the entrepreneur or principle innovator, to establish the target plans for his innovation team. A capable entrepreneur usually develops his conceptual model diligently by grasping the dominant issues of the technological and marketing aspects of the development program. In one form or another, he uses the parameter analysis approach to assemble his building blocks in an orderly fashion according to his own logic.

Ideally, the investor or sponsor (as in a large corporation), while authorizing the investment, should have full understanding of the conceptual model and its range of possible variations during the ensuing development stage. After the investment is made, the investor or sponsor should play only an advisory role by letting the principal innovator run his own show. Overinvolvement of investor or sponsor may produce the detrimental effects described in Chapter 4.

Conventional interface between entrepreneur and investor exists through a business plan which outlines the objectives, amount of capital needed, and includes a detailed chart of the projected cash flow. The Securities and Exchange Commission (S.E.C.) requires that the description of a new venture for public offering include a full disclosure of all foreseeable risk factors. A well prepared prospectus (Red Herring) or business plan should match the innovation measure and the risk factor in a detailed, conceptual model. However, it is also true that in many situations, the innovator has to change his plan after the project is started; in fact, those who can change the plan after a false start are the smart ones. It was for this reason that the noted venture investment expert, Dr. Al Kelley, once commented that he would only invest in one's ability and quality, not in an innovative idea alone. The above statement could be interpreted as a dramatic emphasis on the statistical nature of the risk factor in innovation development and the important role played by the principal innovator, illustrated graphically in Figure 10.1.

Solid-line curve I represents the hypothetical normal risk factor of investment involved in developing a new product. As a rule, the process begins at station O with a very high risk factor. A diligent innovator may significantly reduce the risk factor by prudent development skills to that show at O^+, so that meaningful negotiation for financial support may take place. (The use of stations O and O^+ follows notation in Figure 4.4.)

At point B, income from the product breaks even against cost, and at B^+ all investment should be recovered. (stations B and B^+ also follow notation in Figure 4.4.) From there on, any further investment in that product would have a risk factor of unity, by definition. (The risk fac-

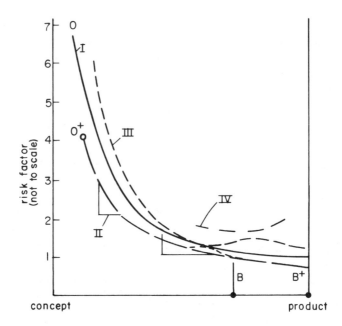

Fig. 10.1 Risk factor of various types of enterprises.

tor involved in developing a single product is different from the risk factor associated with a company, which usually has several products in
different development stages.)

The staircase profile of line II represents the staggered finance program, where risk factors of all investment are reduced by steps every
time new investments are injected into the system. The ability of the innovator to accomplish the set goal at each stage of development in order
to enhance the value of previous investment constitutes the basic concept underlying investment practice.

Curve III represents the speculative risk factor in new product development assumed by industrial firms which have profitable ongoing
product lines but are weak in technological innovation. This difficulty
is characterized in Chapter 4 by the hysteresis loop phenomena for
each individual case implying the higher than normal risk factor on a
collective basis. However, once development passes the halfway mark,
further development can be quite effective, because, from this point on,
experience in professional management is relied upon to assure success
with a better than normal risk. For this reason, many large firms can af-

ford some inefficiency in the earlier phase of product development, even though it is to their advantage to cut development cost at that interval.

Curve IV represents the risk factor for start-up companies which are initially strong due to the skill and perseverance of the innovator; unfortunately, many fail after their product starts to show profit, attributable to the rising risk factor beyond point B. Some plausible reasons for this phenomenon are listed below:

- the innovator has the technological talent but lacks professional management skills in production, marketing, and financing;
- the entrepreneur becomes relaxed and begins to reap the harvest too soon;
- the entrepreneur becomes too ambitious and launches expansion programs too soon;
- the entrepreneur does not appreciate the risk factor of different kinds of capital inputs and relies too heavily upon less costly (in terms of equity), but more risky, capital, such as bank loans and credit financing.

2. EXAMPLES OF THE HIGH RISK FACTOR IN A NEW COMPANY APPROACHING MATURITY

The situation outlined above is a fairly common phenomenon among start-ups, but the sad situation in one particular company (for discussion purposes called XYZ) which was forced into "Chapter 11" (SEC regulation for voluntary bankruptcy) recently (March 1978) is a case in point.

This company manufactured a line of medical diagnostic equipment which sold at 5,000 dollars/unit, each consuming a quantity of throwaway sensors to give the company a sizeable replacement market. By 1977, after five years of struggling, it reached an annual sales volume of 4 million dollars, with marginal profit. The husband (president and technical director) and wife (marketing manager) team gave an impressive lecture about their history and future plans sometime during that year. At that time, the husband described the plan for future Research and Development programs with a 7x5 matrix to show the various branches they could move into systematically. The wife, with nurse's training and some experience in accounting and business management through her previous job as assistant to a professor of management, described her method for developing widespread distributorships to be supervised by a team of company trained salesmen. At that time, there were about two dozen salesmen and numerous dealerships to cover U.S. and overseas markets. The operation appeared to be most impressive, and the fact that it was run by a husband and wife team, with five young children, was certainly incredible. It seems to serve as a rebuttal to the statement by several other entrepreneurs that to be successful in such undertakings, one must sacrifice some degree of family harmony.

The author was also impressed by the rather sophisticated research he learned about in a conversation with their chief scientist, who was preparing a paper to be presented to a physics society dealing with test results generated in their impressive laboratory. Unfortunately, while XYZ company had a saleable product, the grandiose expansion program and image building (the scientific paper) was funded by a loan from a large Boston Bank, with the understanding that the company could get unlimited capital with its assets as collateral. Beginning in the summer, however, the bank cut back on its loans, step by step, from about ½ million dollars to ¼ million, finally demanding balance on the loan. The XYZ company was forced to sell its overseas rights to clear the bank loan and played a payment delaying game with their vendors thereafter, until the accumulated debt reached a million dollars, in addition to a sizeable amount in unpaid taxes. Bankruptcy proceedings were then forced upon them by the creditor committee.

3. THE INNOVATION EVALUATION PROCESS

The failure of XYZ company was clearly the result of poor management, due to inadequately assessing the risk factor of the operation. At the time of expansion, it undertook the following programs, simultaneously:

- doubling the working area;
- expanding the sales force by 50%;
- redesigning the circuit from a five-printed circuit board construction to one of two P.C. boards utilizing the microprocessor (unfinished);
- engaging in image-building scientific research;
- indulging in elaborate fringe benefits for executives (high salary and rented Mercedes).

Had all the expansion programs worked out, management would have deserved fringe benefits, but, unfortunately, all these measures, involving rather high risk factors, failed because they did not use the "perturbation" approach as a lead-in to any major undertaking, as practiced by Wang and Stata in Chapter 4. The worst combination is, of course, using high risk capital (loan or credit) to finance high risk projects. Under different circumstances, loss to the investors would not necessarily damage the principals, who would simply learn from the lesson. Thus, it is very important for investors to have a full understanding of the risk factor involved in various projects to be launched by the entrepreneur. All this cannot be described adequately in a business plan or the risk factors outlined in a prospectus. A more comprehensive way of evaluating and financing an ongoing venture is presented in Figure 10.2

In essence, Figure 10.2 emphasizes the few key parameters which gov-

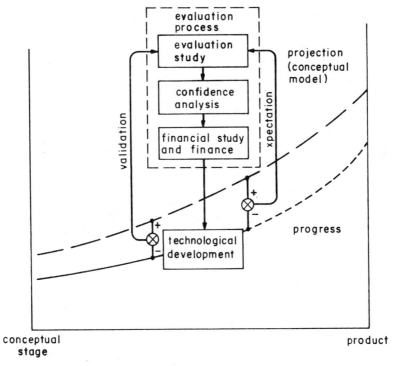

Fig. 10.2. Basic functions of innovation evaluation.

ern the "motivating force" for advancing the status of innovation from concept to fruition. The upper curve is identified as the "projection" (the conceptual model of Figure 4.8 in Chapter 4) and represents the prime focus of the program, even though it is frequently modified according to "need" in relationship to the market condition and the competitor's position. At each stage of development, the evaluation process includes:

1. verifying the projection based on current marketing status and technological feasibility;
2. validating actual progress of the project in comparison with the projection;
3. establishing the expected progress to be made in the next stage of development;
4. examining the confidence factor including:
   - studying the risk factor of the projection;
   - examining the difference between progress made during previous development stage and the current stage;
   - assessing stability of the operation;
   - evaluating the team's capability;
   - establishing the methodology for development;
5. financial support requires attention to:
   - the risk factor which the new capital faces;
   - the potential reward for the new capital;
   - the risk factor inflicted upon existing investment;
   - the reduction of risk factor to existing investment by new capital;
   - various forms and distribution of new capital, such as new stock issue, bank loan, merger, credit buying, SBIC, government grants, etc.

The primary catalyst of the operation resides in technological development, which is responsible for making tangible progress and thereby reducing the risk factor and increasing the confidence factor ("confidence factor" relates to the past accomplishments, while "risk factor" represents the uncertainty of the future).

Philosophically, the diagram in Figure 10.2 is quite similar to the navigation guidance and control operation of a space vehicle, where projection becomes the trajectory plan, the evaluation process corresponds to navigation, and technological development represents control logic. From this standpoint, one may speculate that control logic and optimum trajectory analysis for space travel may be applicable to the development of a new product. The major difficulty for a business, compared to space travel, is that in a scientific operation, most of the parameters are fairly specific, while in a commercial operation, the values of the parameters are assessed differently by a number of individuals with independent perspectives.

4. ANALYSIS OF THE XYZ COMPANY'S REFINANCE CONDITIONS

The bankrupted XYZ medical equipment company, cited earlier, could be a candidate for reinvestment by others, such as its competitors, vendors, distributors, or those with some knowledge of that type of operation. In order to examine the financial viability of a company at this juncture, several parameters are of interest: Charts 10.1 and 10.2 show XYZ company's balance sheet for a six-month period and an income statement for the period ending June 30, 1977, which represents the transition period of its expansion program. The bottom line of the income statement shows the decline of net profit from 7.8% for the six-month period ending June 30, 1976, to 2.9% ending June 30, 1977 and plunging to −16% for the month of June 1977.

During the same period, sales cost increased from 20.7% to 33.3% and then to 45.9%, and new product development increased from 5.3%, to 7.6% to 12.7%. XYZ's bank loan peaked on June 30 at 558,218 dollars; in the next three months, the company was forced to clean up the bank loan by slowing payments to vendors, selling export rights for 225,000 dollars, and borrowing 50,000 dollars in private loans. But all these measurers simply aggravated the situation: The original force fed sales started to drop, despite the still large sales force; shipments were delayed due to difficulty in getting parts from vendors. The operational loss grew continuously, and by January 1978 the stockholders equity was a minus 500,000 dollars, with debts to vendors totalling about 900,000 dollars, employee back pay 20,000 dollars, legal fees for bankruptcy roughly 50,000 dollars (first priority for payment), and back taxes of about 180,000 dollars (second priority for payment).

After trimming away all frills, the management team still hoped to make a turn-around by March. For this reason, they refused an offer from a venture firm to take over the company with 300,000 dollars

Chart I

LIFE SUPPORT EQUIPMENT CORP.
Consolidated Balance Sheet, June 30, 1977
(Unaudited)

ASSETS	JUNE 30, 1977	DEC. 31, 1976
CURRENT ASSETS		
Cash	$ 40,718	$ 102,240
Accounts Receivable Trade, less allowance for doubtful accounts of $17,185	868,151	505,163
Inventories	649,642	363,634
Prepaid Expenses and Other Current Assets	72,020	77,068
TOTAL CURRENT ASSETS	$1,630,531	$1,048,105
PROPERTY AND EQUIPMENT AT COST		
Machinery and Equipment	$ 177,844	$ 133,161
Motor Vehicle	6,645	4,280
Leasehold Improvements	42,987	34,545
Demonstration Equipment	47,059	43,620
Furniture and Fixtures	54,625	49,115
Computer Equipment	36,459	
	$ 365,619	$ 264,721
Less Accumulated Depreciation	(121,808)	(90,846)
	$ 243,811	$ 173,875
Patents Costs	25,239	25,473
Deferred Costs	33,750	37,500
Deferred State Income Taxes	7,757	7,757
TOTAL ASSETS	$1,941,088	$1,292,715

LIABILITIES	JUNE 30, 1977	DEC. 31, 1976
CURRENT LIABILITIES		
Notes Payable to Bank	$ 558,218	$ 373,192
Current Maturities of Long-Term Debt	25,642	6,717
Trade Accounts Payable	791,222	506,663
Accrued Payroll Taxes & Other Expenses	176,973	104,618
TOTAL CURRENT LIABILITIES	$1,552,055	$ 991,190
Long-Term Debt	97,825	69,070
Convertible Subordinates Debentures	54,000	54,000
TOTAL LIABILITIES	$1,703,880	$1,114,260
STOCKHOLDERS EQUITY:		
Common Stock	$ 3,994	$ 3,994
Capital in Excess of Par Value	521,701	521,701
Accumulated Deficit	(288,487)	(347,240)
TOTAL STOCKHOLDERS EQUITY	237,208	178,455
TOTAL LIABILITIES AND STOCKHOLDERS EQUITY	$1,941,088	$1,292,715

Distribution: Prepared by:
J. Baskin   David Provost
B. Gray   July 20, 1977
J. Gray
State Street Bank

Chart II

XYZ
UNAUDITED INCOME STATEMENT
MONTH ENDED JUNE 30, 1977
AND SIX MONTHS JUNE 30, 1977 AND JUNE 30, 1976

	MONTH ENDED 6/30/77	SIX MONTHS ENDED 6/30/77	SIX MONTHS ENDED 6/30/76
NET SALES	$256,953	$2,051,325	$1,508,483
COST OF SALES	(112,039)	916,205	781,545
GROSS PROFIT	$114,914	$1,135,120	$726,934
NEW PRODUCT DEVELOPMENT	(32,774)	(155,524)	(79,425)
SALES & MARKETING EXPENSES	(117,661)	(682,728)	(312,655)
GENERAL & ADMINISTRATIVE	(33,385)	(194,535)	(187,167)
INCOME (LOSS) FROM OPERATIONS	(38,906)	102,333	147,687
INTEREST EXPENSE	(6,678)	(37,384)	(17,807)
INCOME BEFORE TAXES	(45,584)	64,949	129,880
FEDERAL & STATE INCOME TAXES	22,474	(32,019)	(63,603)
NET PROFIT (LOSS) BEFORE			
EXTRAORDINARY ITEM	(23,110)	32,930	66,277
EXTRAORDINARY ITEM	(18,126)	25,823	50,800
NET PROFIT (LOSS)	$(41,236)	$58,753	$117,077

Prepared by:
David Provost
July 20, 1977

Distribution:
J. Baskin
J. Gray
B. Gray
State Street Bank

XYZ
UNAUDITED INCOME STATEMENT
MONTH ENDED JUNE 30, 1977
AND SIX MONTHS JUNE 30, 1977 AND JUNE 30, 1976
RELATIVE TO SALES

	MONTH ENDED 6/30/77	SIX MONTHS ENDED 6/30/77	SIX MONTHS ENDED 6/30/76
NET SALES	100 %	100 %	100 %
COST OF SALES	43.6%	44.6%	51.8%
GROSS PROFIT	56.4%	55.4%	48.2%
NEW PRODUCT DEVELOPMENT	12.7%	7.6%	5.3%
SALES & MARKETING EXPENSES	45.8%	33.3%	20.7%
GENERAL & ADMINISTRATIVE	13.0%	9.5%	12.4%
INCOME (LOSS) FROM OPERATIONS	(15.1%)	5.0%	9.3%
INTEREST EXPENSE	2.6%	1.8%	1.2%
INCOME BEFORE TAXES	(17.7%)	3.2%	8.6%
FEDERAL & STATE INCOME TAXES	(0.7%)	1.5%	4.2%
NET PROFIT (LOSS) BEFORE			
EXTRAORDINARY ITEM	(9.0%)	1.7%	4.4%
EXTRAORDINARY ITEM	(7.0%)	1.2%	3.4%
NET PROFIT (LOSS)	(16.0%)	2.9%	7.8%

Prepared by:
David Provost
June 20, 1977

Distribution:
J. Baskin
J. Gray
B. Gray

cash, on the condition that the management team be changed. The danger was, however, that if the company deteriorated further, it would slip from *voluntary* bankruptcy to *involuntary* bankruptcy at which time XYZ might not get even enough cash from sale of assets to pay legal fees and back taxes. In such an event, officers of the company would have to assume personal responsibility for unpaid taxes.

From the buyer's point of view, it is necessary to take a very conservative position, because his risk factor would be even greater due to inexperience in the market. The buyer must answer the following questions:

1. What is the market's potential?
2. What is the uniqueness of the equipment which will permit him to establish the market, and how long can that last?
3. How can selling cost be reduced (the dominating issue)?
4. How much more R&D work is needed to make the product more economical to produce?
5. Who should run the business, and what should be the incentive arrangement? Should the original team be used under the new arrangement?
6. What is the tangible value of assets?

The example of XYZ company illustrates the high risk factor in new, innovative companies. It is not very appetizing but represents a set of circumstances that an upward bound entrepreneur must avoid instinctively. In innovation and entrepreneurship, this instinct is equivalent to the automobile brake system, which is even more important than a powerful engine.

5. THE ENERGY RELATED INVENTION EVALUATION AND DEVELOPMENT PROGRAM

While the failure of many overextended entrepreneurs tends to raise the risk factor for newly marketed products (described in Figure 10.2) and increase the possibility of bankruptcy, many unsuccessful attempts in innovative development go down the drain unnoticed in the early stages. Many of these involve ineffective use of manpower, even though, in terms of hard currency, the amount thus spent is rather insignificant compared to expenditures of big companies for exploratory, innovation development. Generally speaking, private innovators, while lacking support, equipment, and experience, usually compensate well for their deficiency by their dedication, and, accordingly, once they are able to pass the point represented by O^+, the risk factor for further advance is, generally better than average, as illustrated in Figure 10.1.

The tight money environment in recent years put great restraint on many private innovations. For innovation in ordinary merchandise, this may be just as well, because the restraint can be viewed as a healthy selection process; however, for energy related invention, there is concern that some innovations with hidden merit, obscured by clusters of detracting obstacles, may be lost through the conventional elimination process. To avoid this unfortunate loss and stimulate the general public to participate in the search for energy related innovations, the government established the Federal Nonnuclear Energy Research and Development Act of 1974 (Public Law 93-577) and set up a comprehensive national program for research and development of all potentially beneficial energy sources and utilization technologies. This program is conducted by the Department of Energy (DOE).

To help DOE carry out its responsibility, the Act directs the National Bureau of Standards (NBS) to evaluate all promising energy related inventions, particularly those submitted by inventors and small companies for the purpose of obtaining direct grants for their development from DOE. NBS has established an Office of Energy Related Inventions (OERI) to accomplish this task and has prepared a brochure and booklet explaining the guidelines of the program. In summary, the brochure states that an invention which qualifies for government support should, according to an expert:

1. do what the inventor says it will;
2. have a market once it is produced;
3. offer energy-saving with significant potential benefit to the national energy budget.

6. THE KEY ISSUES IN THE ENERGY RELATED INVENTION EVALUATION PROGRAM

Figure 10.3 illustrates the author's concept of DOE's—NBS Energy Related Invention Evaluation Program evaluation process. In many re-

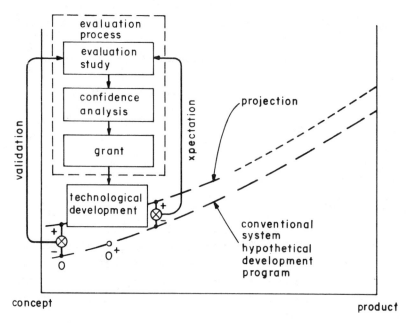

Figure 10.3. Elements involved in government invention evaluation process.

spects, this diagram is similar to the one in Figure 10.2, with some minor variations, such as:

1. Progress in Figure 10.2 now becomes a conventional system with a hypothetical development program. Figure 10.2 shows that, for an ongoing program, it is desirable to establish a projection (or conceptual model as defined in Figure 4.8) to make a comparative analysis of progress (target plan of Figure 4.8). As a rule, progress usually lags behind the projection.

In Figure 10.3, the projection claimed by the inventor is compared with a conventional system; in order for the invention to be worthwhile, it should have a distinct advantage over the conventional system. The curve for the conventional system is constructed as if it were a hypothetical development program, meaning that the new system should be compared to the conventional system at similar development stages and under similar application conditions. For example, when a new carburetor is designed, evaluation and development stages may involve conceptual analysis, pilot model bench test, low cost engine test, high performance engine test, etc.

2. "Validation" in Figure 10.3 is different from validation in Figure 10.2, because the items to be validated are different.

This validation covers the first condition defined in the N.B.S. brochure "it will work the way the inventor says it will" but is somewhat more specific. Many inventors make claims without clearly specifying the proper "conventional system" their invention is trying to replace, nor do they clearly state the objective in terms of specifications.

For example, in proposing a sea wave power generator, it is not enough to simply claim that it can produce power. The specifications that accompany a power generator are quite involved. To help the evaluator and inventor to define more clearly what each kind of energy generating or consuming system should do, an Energy Related Handbook has been proposed by the M.I.T. innovation center.

One important aspect of validation is examining the adverse effects which always exist with the advantageous, novel properties playing the primary role. Conversely, the value of some novel innovations may be obscured by apparent adverse effects. In either case, it is desirable to discover the adverse effects as early as possible, so that eliminating them can play a dominant role in ensuing technological development.

3. Showing the evaluation process (broken line) at the left of the diagram, instead of in the middle as in Figure 10.2, emphasizes the vulnerability of innovation and its remoteness from the finished product, which is represented at the right.

In a large company, it is quite common to make an extensive market analysis of a new product before the technology is firm. For example, in the development of the C.A.T. scanner, Analogic Inc., (Prologue) made an artist's rendition of the finished machine before the technological scheme was drafted, because, for such an application, the external configuration provides the link between the computer interface innovator and the overall system designer. For energy related inventions, a bridge to the future should be established through improvement over existing systems on the basis of technological parameters and attempts should be made to examine the higher echelon of technological parameters to achieve broader understanding.

For example, in evaluating wind power generators, devices for individual rural adaptation and those intended for large scale power distribution require different reference systems for means of comparison. Accordingly, in evaluating the cable windmill (Chapter 8), the reference system for comparison should be an hydraulic power generating system instead of another form of windmill. By selecting the hydraulic power generating system as the reference, the need for a reservoir for energy storage becomes apparent. In the initial evaluation, efforts should be directed toward the geographical study of wind power distribution in relation to the topographical conditions in various regions. (Such an analysis is similar to the hydrographic study done prior to designing hydraulic power stations.) In evaluating cable windmill, the structure configuration and drive system were determined to be the basic parameters of such a power generation device. On the other hand, the design of the airfoil was not particularly emphasized, because, in this adaptation, the configuration of the airfoil is not the key issue.

The cable windmill was proposed with the assumption that in some valleys, a steady wind prevails; without this initial requirements, refining structural design is useless. Likewise, premature development of the airfoil tends to overshadow the examination of some important, hidden problems associated with structural design or terrain.

Recognizing the hierarchy of various technological parameters is considered a part of parameter analysis methodology.

7. EVALUATING FINANCIAL SUPPORT FOR ENERGY RELATED INVENTIONS

Expectation

The purpose of financial support for innovation development is an effort to decrease the risk factor in the most efficient manner and at a clearly defined stage, identified as "expectation" in Figure 10.3. Quite often, the risk factor exists in some area clearly recognized during validation phases and is followed by a speculative remedy—later development concentrating on that area which verified assures remedy would constitute reducing the risk factor. For example, one particular invention attempted to use centrifugal force to separate the dozens of ingredients in automobile exhaust, then tried to recover some of the unburnt hydrocarbon, while using a scrubbing scheme to remove pollutants. After coupling a crude testing device with an automobile, the inventor claimed a small percentage of improvement in gasoline mileage. Apparently, his test results were neither convincing enough to sell the scheme to automobile component companies nor to raise enough capital to start his own company. With limited government support, the route this inventor should take to minimize the risk factor represents a challenging technological decision.

After making a parameter analysis, it was suggested that he examine the mechanism used in separating exhaust components. For his next experiment, it might not be advantageous to study the performance of the total system, including the automobile's many difficult-to-define factors (carburetor condition, load condition, etc.). On the other hand, if the dominant parameters of the separation mechanism are well understood and supported by laboratory evidence, their application to the automobile or other systems becomes more obvious and constitutes a reduced risk factor. Doing so also means that the "system" used for comparison, as shown in Figure 10.3, is no longer an automobile, but a conventional, gas separation system.

Confidence Analysis

The comment made by Dr. A. Kelley, cited earlier, that he would invest in a man; based upon his entrepreneurial skill and not in his particular invention, should be qualified on two counts, before used to examine the confidence factor in government funding for energy related inventions:

- The energy related invention is not to be funded to generate new enterprise.
- The government is not concerned with a return on investment to the same degree as the private investor.

Thus qualified, the government's position regarding confidence in the inventor's ability should be dircted toward providing reinforcement instead of expresssing reservations. In this respect, a cooperative R&D program, using university facilities and guidance to help individual innovators examine key parameters and isolate pertinent issues for conducting experiments, would be an advantageous auxiliary.

This chapter represents a conceptual outline of an Invention Evaluation Manual to be developed by the author under contract with the National Bureau of Standards.

11
PARAMETER ANALYSIS
AND STUDENT
RESPONSE

1. INTRODUCTION

At M.I.T., there are usually various types of contests to tax students' innovative talent and competitive spirit. None, including two prize winning student invention contests (sponsored by the innovation center) can match the popularity achieved by the "Design Project," adjunct to the "Machine Design" (M.I.T. subject 2.70) conducted by Professor Woodie Flowers of the mechanical engineering department and a member of the innovation education council. This project parallels the lecture portion of the course and occupies the first half of the semester. At the beginning of the semester, each student is given a "brown paper bag" containing a collection of "goodies," such as, a few sheets of plastic material, pieces of bolts, nuts, etc., and an instruction sheet. Each student is instructed to use the material from the bag to build a device which meets the specifications outlined in the instruction sheet.

The contest is based on the process of elimination as in any tournament. The "test ground" (arena) is well engineered and equipped with accurate measuring and actuating contraptions to avoid any confusion and to allow a faster turnover, so that it takes an average of five minutes to make a decision; thus, in one spirit-charged hour, the winner quickly emerges. In the finals, the huge lecture hall fills up with several hundred spectators rallying for the contestants as in a boxing match. Dr. Land, who had donated 150 minature motors for the 1977 contest, claimed that he had not seen so much excitement for quite a while. Enrollment in this class has increased over the past five years from below 50 to about 140. This kind of contest serves well as a model for teaching innovation, or similar subjects, dependent on the student's own motivation. A few observations regarding these subjects are:

- The instructor (Woodie Flowers) has been very innovative in designing the contest (which changes every year), so that the task matches well the capabilities of the students and yet is sufficiently fascinating to arouse intellectual interest.
- The goodies selected to fill the brown bag are very important: The bag must contain enough ingredients to permit normal design and yet offer sufficient variation to stimulate innovative ideas.
- The design and construction of the test apparatus must be well executed, so that fast and accurate test results give the student confidence in a meaningful outcome.

The author, invited by Flowers to participate as a contestant, was amazed at the amount of thought and effort one can put into preparing for the contest and the infinite variations that can be generated from material in the brown paper bag.

2. PARTICIPATION IN "THE POTENTIALLY GREAT RACE"

At one of the weekly innovation education council meetings, Flowers gave.Y.T. Li a "brown paper bag" and invited him to participate as a "phantom contestant" in that year's student design contest, billed as a "Potentially Great Race." At first, Y.T. Li declined on the ground that there was no "market" for the product thus created. The following week, Flowers brought over a second bag, saying that he realized that Professor Li deserved two bags to allow for mistakes and added that Li need not feel embarrassed if something went wrong, because Flowers himself had already set the failure precedent by participating the year before with a cart that flipped over halfway down the test track.

Reluctantly, Li promised to think about it. Not long after, he was persuaded to participate in earnest by his son Kenneth, who was a contestant. Thus, on the night before the contest, Kenneth went home to work in the basement, while Li stayed at M.I.T. until 3:00 A.M. to add the finishing touches.

3. INSTRUCTIONS FOR THE CONTEST

Object

To design and build a device which travels down a given track (see Figure 11.1), turns off a light bulb by interrupting a beam, reverses direction, and travels back across the "start" line as quickly as possible.

Constraints

Material

Except for glues and adhesives (used only for bonding) and nonfunctional decorations, the devices must be constructed entirely from materials supplied in the kit.

Energy Source

The energy used by the device to accomplish the objective must come solely from a change in the altitude of the center of gravity of the device.

Size

The dimensions of the device must always allow that it be entirely contained by a 2-foot cube which has its base in the same plane as the base of the device.

Time

The devices will be considered to be complete at 12:00 noon, October 22, 1974.

Evaluation

A winner will be chosen in a single elimination tournament. Races between pairs of PR's (Potential Racers) will be run on parallel tracks.

The winner of each race will be the PR that is first to make a complete return across the starting line after having turned out the light. (Grudge matches will be run as time allows).

Details:

1. During competition, after a racer is called to start, a maximum set-up time of one minute will be allowed.
2. Kit materials may not be changed chemically.
3. Each PR must properly interface with the starting pin on the track.
4. A drawing of lots will determine first round competitors.
5. After the runoffs, the surviving PR's will be impounded and will be released 15 minutes before the finals begin.
6. After each PR's initial race in the runoffs, no major design changes will be allowed.
7. No manipulation of, or interaction with, a PR will be allowed while it is racing.
8. The tracks will be placed on tables—competitors will be responsible for saving their PR's, should they stray off the track.
9. The devices may contact only the top surface of the track.

Kit of Materials for "A Potentially Great Race":

1. 1 ft × 1 ft masonite sheet
2. 11 in. × 14 in. × .010 in. celluloid sheet
3. two white cardboard tubes, one with OD slightly larger than the other
4. two white plastic containers
5. 1 ft-strip of metal banding material
6. 1 ft-wooden strip
7. 1 ft × ¼ in. wooden dowel
8. one interdepartmental envelope containing a length of lycra thread
9. 1 ft-welding rod
10. one piece of string approximately 10 ft long
11. one pencil
12. one paper sack containing approximately 1 lb of sand
13. 3 in. × 3 in. × ¼ in. piece of plexiglass
14. two pieces of polyflow tubing

15. three caster wheels
16. four rubber bands
17. four paper clips
18. one in. steel rod
19. four 5 in. × 7 in. note cards
20. four 8½ in. × 11 in. sheets of paper

4. PARAMETER ANALYSIS OF "THE POTENTIALLY GREAT RACE"

As a Phantom contestant, Li did not participate in person but sent a student, Marco Flores, from the invention class to represent him and make the demonstration test to the 2.70 class. The test went smoothly with an elapsed time of 4.2 seconds against the 7-second best time achieved by the other contestants. The major design difference was that most of the students took the easy route by using the weight of the sand to drive the wheel directly until the cart reached the end of the track, then reversed the winding direction of the string on the wheel shaft to drive the cart backward. In this manner, no kinetic energy recovered that might still remain at the end of the outgoing pass. The model built by Li utilized the concept of a "conservation system," similar to that of a pendulum.

A parameter analysis was then written by Li to be distributed to both the machine design and invention classes as a reminder that it pays to examine the root of the problem (the higher echelon of parameters) instead of plunging into the final configuration by instinct. In this example, instinct is designing a cart with a simple mechanism to drive it forward and then reversing it at a predetermined distance. The higher echelon of parameters includes consideration of energy and acceleration, the trajectory of the center of mass, the coefficient of the friction of the wheel on the track, etc. Some of the highlights of the parameter analysis include:

- The role played by the energy source in driving the device:

  The energy source in this system is the mass of the sand provided in the brown paper bag. The available energy potential is achieved by lifting the sand to the maximum allowable height of 2 feet.

  Energy consumption results from the loss of friction as the device runs over the track. With the use of wheels, it is assumed that fric-

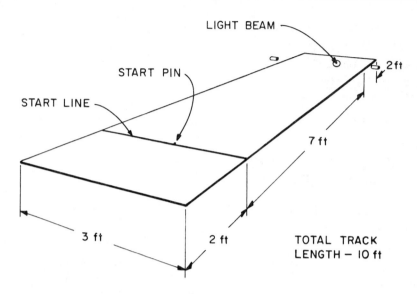

TRACK SURFACE – PAINTED PLYWOOD

LIGHT BEAM – 3/4" INCH ABOVE TRACK, 7 ft FROM
            START LINE

START PIN – RETRACTS BELOW TRACK SURFACE
            FOR START

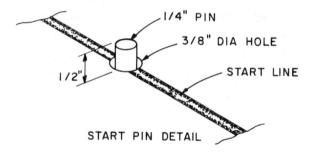

START PIN DETAIL

Fig. 11.1. A potentially great racetrack.

tion loss can be minimized through good design practice and that the cart can be designed with a low weight penalty.

- The goal of the contest is to minimize travelling time, which suggests high velocity in the horizontal direction, obtainable by integrating acceleration.

A pendulum serves as a good model to examine the *trajectory* of a conservative system. A pendulum with a swing of 7 feet and a head of 2 feet has a length of 4.35' and a period of: (see Figure 11.2)

$$2\pi \sqrt{\frac{4.35}{32.2}} = 2.31 \text{ sec.}$$

In actual test, it is about 2.5 seconds. Before considering a practical scheme to implement the given problem with all its constraints, let us give some thought to the pendulum: The string of the pendulum provides a circular arc for the trajectory of the pointed mass. One may ask, is the circular arc the best trajectory for minimum time? For instance, one can build a wire-guided pendulum like the one shown in Figures 11.3a and 11.3b. Intuition in all these examples illustrates that it is desirable to accelerate as much as possible at the beginning of the trajectory and decelerate as much as possible at the end, so that a fast velocity occurs in the middle.

- One possible configuration is a simple vehicle in the form of a 2-foot diameter cart wheel with the weight attached to the rim of the wheel so that with a little trick in designing the trigger system

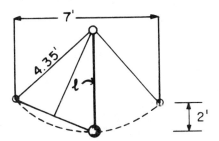

Fig. 11.2. Model pendulum.

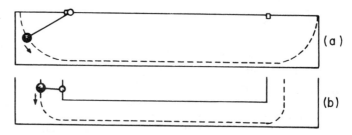

Fig. 11.3. Wire guided pendulum.

(a complete revolution of a 2-foot wheel is just short of 7 feet), it covers the 7-foot track as shown in Figure 11.4.

The trouble with this scheme is that it has a very slow horizontal veloctiy at the middle span. To state this in another way, it has a very slow horizontal acceleration at the two ends of the path.

- Conceptually, the objective is to design a cart which carries the driving weight (sand) so that as the cart moves from one end of the 7-foot track to the other, the driving weight will describe a trajectory similar to that of a pendulum with an approximate 2-foot drop in height at the middle.

- The three positions (*a*, *b*, *c*) of the cart (broken lines) and the trajectory of the driving weight (dash-dot line) are shown in Figure 11.5.

- The relative trajectory between the driving weight and the cart is shown in Figure 11.6

The design objective is to couple the driving weight (sand box) with the wheel of the cart so that while the weight is moving along its trajectory in relation to the cart, the wheel will turn as prescribed by the desired relative position of the cart with respect to the track.

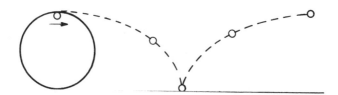

Fig. 11.4. Tumbling wheel with an eccentric mass center.

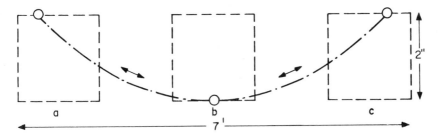

Fig. 11.5. Trajectory analysis of the mass center of the conservative potential racer.

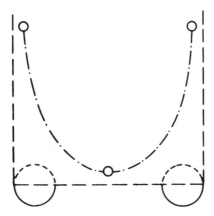

Fig. 11.6. The "U" shaped trajectory of the mass center with respect to the envelope of the cart.

- The mechanism needed to guide the trajectory of the weight depends upon what is contained in the brown paper bag, what is available in the designer's bag of tricks, and the degree of experience in matching them through frequent exercise.

  In this instance, a two-member linkage was chosen to support the weight and to generate the U-shaped trajectory. The simple geometry dictates that for a swing of an angle θ between members a and b, member b should swing at angle α with respect to the center line of the contraption, as shown in Figure 11.7. If θ and α are to move in a fixed ratio, it can be accomplished with a pair of pulleys (a segment is needed for the fixed pulley) and coupled together with a string as shown in Figure 11.7.

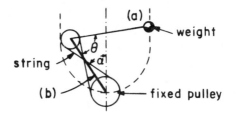

Fig. 11.7. Two member linkage.

- The finished configuration is shown in Figure 11.8, where a great deal of detailed design "tricks" were used to take full advantage of what was available. A relief mechanism was also added so that in the return pass the weight would not be re-elevated in order to allow the cart to coast back freely.
- Final adjustments and testing of the contraption were undertaken by Marco Flores. One significant contribution made by him was the use of graphite from the pencil provided in the brown paper bag as a dry lubricant. Final adjustments necessitated many trial runs and adjustments to the wheel pulley size and surface condition to attain traction and distance. The work also involved perfecting the starting and finishing trigger mechanism.

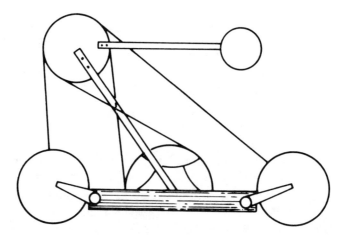

Fig. 11.8. Overall assembly of the conservative potential racer.

The total effort that went into designing, constructing, and testing this conservative potential racer was certainly much more than could be expected from a student to satisfy course requirements. The effort put in by Marco Flores alone was probably comparable to that of a regular contestant. But the effort appeared to be well justified two years later when another student (Susan Kayton) requested that the author do a parameter analysis for that year's 2.70 project (Fall 1976), "The Sandbox Derby." She stated that it would not be fair to ask the author to participate in the contest again but that a few of her classmates would appreciate his assistance in, once again, making a parameter analysis. The author promised that it would be done after the contest and was most delighted to watch Kayton take first prize a few days later. Appreciation of such a methodology by a capable student is, of course, the greatest encouragement a teacher can receive. Thus, even the parameter analysis of the Sandbox Derby, done as promised, would not add much to the present discussion.

In retrospect, when the author refused to participate in The Potentially Great Race on the grounds that the product did not have a market, he was totally wrong. Woodie Flowers' "product" was not only highly marketable at M.I.T. but was copied by many other schools. On the basis of his ingenuity in designing the new contest every year, his course will flourish, with or without parameter analysis, because the sheer fun and excitement will draw students. On the other hand, parameter analysis is a methodology which needs a vehicle to carry it, equivalent to an invention that is meaningful only when a "need" is perceived first.

12
TEACHING TECHNOLOGICAL INNOVATION TO SERVE INDUSTRY

1. BACKGROUND

Over the past ten years, the need for stimulating technological innovation has been frequently discussed in government, industry, and universities. Various schemes for training innovators and entrepreneurs were attempted at many universities at the time when the National Science Foundation decided to spur this movement with substantial funding of several innovation centers; M.I.T., The University of Oregon, and Carnegie-Mellon University were finally chosen to carry out the five-year experiment with independent academic programs.

At M.I.T., faculty members from five departments of the School of Engineering and some from the School of Management organized the innovation center, which was structured into two branches: one was identified as the innovation education council, responsible for class-room teaching in innovation related subjects, while the other was identified as the innovation co-op, in charge of organizing new product development projects. Over this period, about 500 students took various subjects, and about 50 students were involved in various product development projects. About 19 patents were filed, 11 were issued, 80,000 dollars in royalty payments has been received, and four companies have been organized involving students in the exploration of products they developed.

While most of the students found the innovation center experience very valuable, it did not achieve a self-supporting status after five years of operation. (An arbitrarily set goal to indicate that the funding was committed for five years only.) Being self-supporting is a commerical goal, which is usually carried out in the free enterprise environment with no additional burdens above those required for making profit. On the other hand, the innovation center experiment was structured primarily for education, with the classroom teaching portion subsidized totally by the center over the entire five years. The co-op also existed primarily to counsel students who desired to become entrepreneurs. Thus, the primary objective for this experiment was studying the interface between university and industry and exploring the appropriate mode of training entrepreneurs and future industrial leaders, while providing a service to industry in developing innovative products. At present, the decision whether or not a university affiliated innovation center is justified hinges upon the following concerns:

- Need: Technological innovation, at one time the hallmark of U.S. industry, was intuitive in nature and seemed to be self perpetuating; how critical is the need for innovation now that some new stimulation must be undertaken?
- Plausible measures for obtaining quick results: Infusion of positive stimulants, such as tax abatements for the innovation process, tax relief for capital investment, extension periods to carry over tax losses, research and development support for small business, government assistance in venture capital financing, and introducing more flexibility into various regulatory laws, are measures which may produce fast results, yet each of these may have inherent, adverse effects; accordingly, are these plausible measures adequate and sufficient to answer the aforementioned need?
- Long-term solution: Developing human resources by training innovators and future leaders while they are still in college may be considered a long-term solution to needed techological innovation; how can this process be used effectively in conjunction with previously mentioned measures for achieving fast results?
- Can innovation be learned?: The questions raised here are to be examined at the symposium for Innovation and Innovation Centers in May 1978 at M.I.T., along with this work. A proposed structure for the innovation center is presented in the last section of this

chapter. However, the characteristics and conditions to enable this proposed center to effectively carry out the aforementioned objectives will be examined first.

2. THE CHARACTERISTICS AND CONDITIONS FOR TRAINING TECHNOLOGICAL INNOVATORS

a. While the goal of technolgical innovation implies new product development aimed at satisfying social needs with tangible financial benefit, the educational function of this undertaking must exist on a solid, academic foundation.

Generally speaking, financial reward is used extensively in industry to provide incentive, but excellence in education requires motivation instead of incentive. While incentive is derived from financial reward, motivation is derived from recognition—especially recognition from those whose esteem is valued.

Skillful industrial management often mixes appropriate form of recognition to motivate its people. After all, incentive causes financial drain, and there is always a diminishing return beyond a certain amount of monetary incentive. Furthermore, only certain activities are amenable to incentives, while many others are more responsive to motivation.

If technological innovation is to be taught in a university, nationally accepted directives must be defined. The building block parameter/parameter analysis methodology introduced in this book is an initial attempt at identifying directives for innovation education. The ultimate recognition of the need to teach innovation will occur when teaching innovation is accepted in academic communities throughout the nation as a normal part of education.

b. New product development must be accompanied by proper financial incentives.

Generally speaking, universities in the U.S. have refrained from pushing the commercialization of technology in earnest and patents licensing is generally limited to fallouts from scientific research. The financial yield from university undertakings has not been too lucrative, and the formula for sharing royalties with inventors has not been attractive either. Some faculties with a marketable invention are willing to carry an extra workload, frequently looking to outside venture capital for support. To some extent, the growth of industry along Route 128 around Boston, or in the Silcon Valley near San Francisco, was due to the exodus of innovative ideas from universities. If it is deemed desirable to pursue product development as an activity pertinent for training innovators, the university must be prepared to embrace similar incentive concepts employed by industry in conjunction with motivation through academic recognition.

The delicate balance between incentive and motivation deserves some careful study. Following this line of reasoning, an experienced entrepreneur, who has savored the fruit of innovation and is ready for new challenges, such as those found in education, would be the ideal candidate for teaching innovation. The question remains, how can a university establish a system that offers such individuals sufficient recognition in order to motivate them to teach technological innovation?

c. Students must be motivated to learn technological innovation.

The success story of Dr. Edwin Land, who left Harvard University as an undergraduate to launch the Polaroid Company, led many people to associate innovation education with developing one's own innovative idea. In reality, success in innovation must result from screening many viable ideas, and it would be misleading to encourage the student to focus exclusively on his own idea, precluding an open mind with regard to the ideas of others. Furthermore, a major task in innovation is accomplished by the entrepreneurial role, which must be cultivated through exposure to and participation in real product development projects (which usually are not his own).

Finally, the best way to motivate students is through projects leading to tangible results accomplished within the normal, academic time unit, usually, three months. The design contest mentioned in Chapter 11 is a typical example of an innovation exercise favored by a majority of students. It would be doubly fruitful if the contest were structured to maintain a similar format, while, at the same time, including some real application.

d. It is important to develop an ideal innovative environment.

The ideal innovative environment is one where a large number of recognizable needs flash by and the cross flow of numerous, technological building blocks or concepts occur. This need for a "reflux" of knowledge to generate technological innovation is analogous to using a large quantity of reflux of the distillate to get a high concentration of a certain ingredient from a distillation tower.

3. A PROPOSED STRUCTURE FOR TEACHING TECHNOLOGICAL INNOVATION

Figure 12.1 shows the conceptual structure of a proposed innovation center designed to train future innovators and industrial leaders. The center is comprised of two major branches, one engaged primarily in classroom teaching and research on innovation methodology, the other with innovation development aimed at generating new products to serve industry and giving students experience in fulfilling that objective.

The parameter analysis approach discussed in Chapter II is considered to be the appropriate methodology to teach students interested in pursuing a career in innovation. The area is identified in Figure 2.3 of Chapter 2 as building block versus technology parameters, which are now included in various engineering curricula. Chapters 6 through 9 represent an initial attempt at identifying the various building block parameters and describing how they may be generalized and expanded into a body of knowledge with adequate structure for efficient teaching.

What has been attempted so far in this work only scratches the surface of this fascinating area. Needless to say, a considerable amount of research will be required to broaden and sharpen this knowledge, requiring the efforts of experts who have the urge to teach, in addition to a considerable amount of industrial experience, especially in technological innovation. Students intent on choosing innovation as a career should be allowed to elect a double major, with course work in innovation, as well as one other discipline in the school of engineering or management. They should have adequate professional training in a few, major areas.

The second branch of the innovation center is the innovation co-op, set at an arm's length from the university, which refers to the fact that

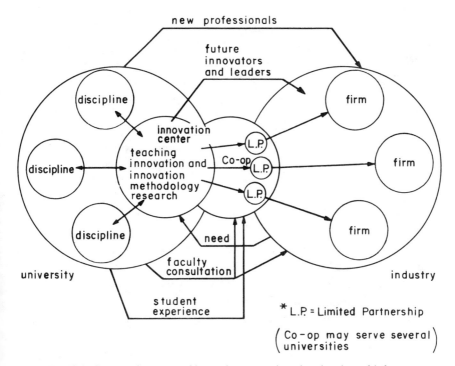

Fig. 12.1. Conceptual structure of innovation center vis a vis university and industry.

the co-op is a nonprofit, charitable organization, affiliated with the university but not subject to its routine, administrative constraints. Accordingly, the co-op can devise its own flexible management policies concerning licensing arrangements, products marketing, and confidential obligations to industry. Of particular importance is the progress-oriented approach to compensation. For example, university faculty may be engaged as consultants and compensated according to their contribution. This would not be feasible if the organization were structured within the university.

The co-op overlaps with the university through its common bond with the teaching branch of the innovation center, and a few endowed professors preside over and direct both teaching and co-op operations. Each of these professors is strong in technology, talented as an innovator, has risen beyond the temptation of financial gain or the tendency to show off superficial skills. Under their guidance, various industrial needs will be examined and innovated, and projects will be dissected, with some designed for student practice in innovation and other's channelled to professionals for development.

As projects emerge showing signs of a quantum jump in performance, further development may be carried out under the "limited partnership" (L.P. in Figure 12.1) arrangement, where activity is primarily commercial and changes from "not for profit" to "for profit," becoming subject to state and federal taxes in the same manner as all other commerical operations. Private investors may participate in these limited partnerships, and, therefore, risk and gain are confined to each group.

Each of the limited partnerships should have one person serving as entrepreneur with the job of bringing a new product to a significant stage of maturity, so that it can be assigned to a new company or licensed to an existing company. There is a general need among many well run industries for new products which they cannot develop themselves at a desirable low risk. Figure 12.2 shows the risk factor curves discussed in the last chapter and reproduced here to illustrate how the innovation co-op, through a limited partnership, may help serve this need. Curve B represents the risk factor for a typical management oriented company in developing a new product; curve A represents the plausible risk factor in new product development when handled under the innovation co-op, limited partnership arrangement.

Along curve A, point a_0 represents a raw new idea or invention generated as fallout from scientific research in the university environment. It was not developed with a new product in mind and, in general, has a very high risk factor. Under rare circumstances, with a unique innovation, such as the magnetic core memory, which came into being at the early stage of a major, technological breakthrough, the inventor and the university may reap a sizeable benefit. If the university's function is limited to licensing inventions from scientific fallout, then the infrequency of payoff would imply a high risk factor, as shown at point a_0. By addressing the market need and stimulating interface between industry and skillful innovators, the risk factor of the inventions thus generated would be lowered to point a_1. If prototypes were developed, as the

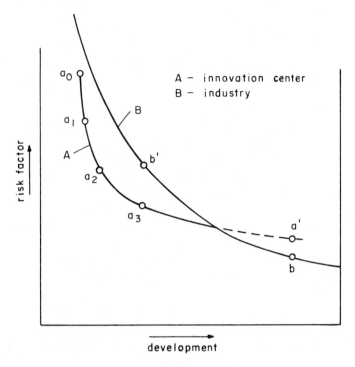

Fig. 12.2. Hypothetical risk factor function of innovation center versus industry.

M.I.T. innovation co-op has done during these past few years under N.S.F. support, the risk factor could be further reduced to point a_2.

Risk factor is a complicated phenomenon. In general, it is the inverse of the return on investment but is not the same as real property, such as a house, whose intrinsic value remains once it has been developed to a certain stage. For example, a patent on a new product can only enjoy its market value for a short period (usually less than the U.S. patent life of 17 years). On the other hand, there is a difference between the risk factor and the market value of a new product, as observed by the potential licensee. In negotiating, it is always better to maintain a flexible latitude to allow room for bargaining; thus, with the structure of the limited partnership, it permits the operation to move from point a_2 to point a_3 and beyond and extends the range of the negotiating position, while the risk factors are continuously reduced. At the time when the limited part-

nership is spun off, it may chose to exist as an independent entity with a
risk factor such as a' or be absorbed by a larger firm with more management experience and broader marketing outlets to reduce the risk factor to point b.

4. THE CHARACTERISTICS OF THE LIMITED PARTNERSHIP

The objective of limited partnership is to make new product development an ongoing activity affiliated with the university incorporating the
following features and considerations:

 a. The output of the operation is a constant stream of small enterprises with a lower than average risk factor.
 b. Paralleling a), there will be a supply of new products developed to
  be licensed to well established industries; both a) and b) should
  help the national economy.
 c. A limited partnership will provide investors in each project with
  clearly defined financial gains to compensate for their risk.
 d. A limited partnership will provide the entrepreneur in charge of
  the project with clearly defined incentives and motivations.
 e. A limited partnership, with its unabashed commercialization objective but total tax commitment, will not affect the university's
  tax exempt status.
 f. The university, while assuming relatively low risk in managing
  the limited partnership, can yet derive significant financial reward
  due to the activated royalty income of the more realistic projects.
  undertaken by the L.P.'s.

5. THE VARIOUS ROLES OF THE INNOVATION CO-OP

The dual function of the innovation co-op is education and business.
Physically, it should have a product development laboratory and support
shops for fabricating various models and prototypes of new products.
These facilities could be built up as needed by the various L.P.'s but retained, in most cases, by the co-op as general purpose equipment to support various L.P.'s and student laboratory exercises.

The ability of the co-op to lower the risk factor in new product development may be attributed to the following three factors:

a. The co-op may be exposed to a large number of industrial and social needs because of its affiliation with a university which has relatively free access to information coming from both government and industry.
b. Innovation taught by the faculty of the innovation center tends to rely heavily on parameter analysis and should yield better innovative ideas.
c. Free access to the wide pool of technological science and business knowledge in the university environment and flexible consulting arrangements with faculty makes product development less hazardous.

The education role of the co-op can be sharpened by incorporating the relatively large influx of innovative ideas and needs into teaching material and laboratory exercises.

Chapter 11 illustrates the excellent student response to innovative exercises with well defined and limited goals. It also appears possible to design student contests which lead to marketable products under the innovation co-op umbrella. In general, students are constrained by their academic schedule and motivated by the challenge of new knowledge so that it is unrealistic to let them be responsible for a major product development project requiring a time commitment beyond what they can afford and demanding more knowledge than they have already acquired. Thus most of the product development work load should be assumed by professionals, while structured to permit the students full exposure to innovation activities.

The operational function of the co-op is, by itself, a challenge because of its dual educational and commercial role. One purpose in having the co-op set at arm's length from the university is to minimize its obligation to the bureaucratic routine of the university, which has a different set of constraints and would impose an excessive burden on the co-op's operation.

EPILOGUE

Mr. Charles Tandy of the Tandy Corporation visited the M.I.T. innovation center on April 4, 1978 to give a lecture on marketing. By coincidence, it was the fifteenth anniversary of his having assumed control of Radio Shack of Boston, a small company suffering an annual loss of close to ½ million dollars. Under Tandy's direction, the trend reversed, and Radio Shack has now become a worldwide chain, enjoying 1 billion dollars in annual sales, while maintaining the leading profit-making position and the highest growth rate among major retail businesses.

One marketing strategy that Tandy discussed with students was development of his company's own identity by product marketing with his own brand name, even though the product was fabricated by others. In so doing, he was not only able to negotiate among several suppliers to get the best price with the best quality control but could also realize the accumulative effect of his own advertising. Being thus established, he was able to exercise his own innovative approach to marketing. For example, in the vacuum tube days, after identifying a low cost supplier, instead of reducing his own retail price as an inducement as his colleagues recommended, he escalated the price 10% above the competition, while offering an unprecedented lifelong guarantee. His own statistical analysis showed that the lifelong guarantee would cost him less than a reduced price, and yet it would carry a much higher incentive for customers. Above all, he achieved "uniqueness" for his product in the face of strong competition in the field. This particular move represented a novel approach and a quantum jump in performance—like an invention—even though it was not a patentable idea. Nevertheless, it was an innovation of the sort that is as amenable to parameter analysis as technological innovation, which we discussed earlier.

Tandy's overall strategy was to apply his own formula of retail business management to a large number of similar retail stores. His formula was to give the store manager a strong incentive geared to profit making and a well-defined task (such as tight inventory control), well trimmed catalogues, etc. He developed this formula for merchandising leather craft and found that it worked equally well when adapted to the electronic product. His stores are in fact his first product line, which he established at a rate faster than one and a half stores per day during the span of several years.

After having established one of the largest retail chains in the world he then moved into manufacturing and made a stunning success recently in the C.B. radio and then in the microcomputer. In the latter case he claimed that it was developed with only a shoestring budget by one of his engineers.

After the emergence of the microprocessor many entrepreneurs and engineers jumped onto the bandwagon of the microcomputers. Many tried and failed similar to what happened to the electronic calculators, T.V. games, and digital watches. Indeed, the spirit of free competition is a game for all and all for the winner. The winner is the one with the strongest combination of technology innovation, market innovation, and team spirit. In the world of free competition, Charles Tandy championed with the genius in marketing innovation and team inspiration.

Thus, while it is the purpose of this book to develop the methodology in technological innovation it is by no means self-sufficient as illustrated by the Saga of Charles Tandy (the author was saddened by Tandy's sudden death September 1978, a day before his appointment with the author).

Whether one may win with technology innovation, marketing innovation or team motivation a balance of all three is paramount and the situation for any of these three to flourish appears to be in the environment of the free competition society. The author is very encouraged to see a change in this direction in China recently with their Four Modernization Policy.

How to adapt the spirit of free competition while embracing the doctrine of a planned economy and public ownership for China is a social innovation which is on an even higher hierarchy than technology and market innovation. However, to enrich the free competition spirit in some form of planned economy is very likely the trend of our spaceship-earth and that is the ultimate challenge to the innovation of human kind.

Index